S0-CAH-159

Aerosol Particle Size Analysis

Good Calibration Practices

Aerosol Particle Size Analysis

Good Calibration Practices

W. D. Griffiths, D. Mark, I. A. Marshall and A. L. Nichols

Aerosol Science Centre, AEA Technology, Harwell, Oxfordshire, UK

A catalogue record for this book is available from the British Library.

ISBN 0-85404-452-3

Published for AEA Technology
by The Royal Society of Chemistry,
Thomas Graham House, Science Park, Milton Road, Cambridge CB4 4WF
For further information see the RSC web site at www.rsc.org

Typeset by Paston Press Ltd., Loddon, Norfolk
Printed and bound by Bookcraft (Bath) Ltd

Preface

Particle size analysis is extremely important in the characterisation of aerosol ensembles. The transport of such particles (and droplets) is determined by a combination of factors, of which a knowledge of the size distribution is particularly important. Hence, equipment used to measure particle (and droplet) diameters and size distributions need to be operated with competence and confidence. A Manual of Good Calibration Practices has therefore been prepared, with guidance from the National Calibration Forum for Aerosol Analysis (NCFAA) established through a Valid Analytical Measurement Programme (VAM14) of the National Measurement System Policy Unit (NMSPU), UK Department of Trade and Industry. The aim has been to prepare a starting document to aid non-specialists in their understanding, valid operation and calibration of aerosol particle-size analysers, so that the various types of equipment are used correctly and traceable quantitative data are obtained with confidence.

Contents

Acknowledgements

The authors would like to thank J S Backhouse and D R Booker for their assistance in the preparation of this manual, and all members of the National Calibration Forum for Aerosol Analysis for their advice in defining the contents.

Production of this report was supported under contract with the UK Department of Trade and Industry as part of the National Measurement System Valid Analytical Measurement Programme.

CHAPTER 1

Introduction

Many of the necessary procedures for the calibration of aerosol instruments are ill-defined and lack proper instructions and documentation. Under normal circumstances, initial calibrations of new instruments are carried out by the instrument manufacturers or their agents, with further calibrations being provided when the instruments are returned to them for cleaning and service. In addition, a very small number of specialist laboratories provide calibration and characterisation services for various aerosol samplers and monitors. However, in a recent survey of user needs (Lewis *et al* (1993)), it was found that 58% of the aerosol instruments used at that time were either calibrated in-house, or not calibrated at all. Clearly this situation is unsatisfactory as the quality of calibration may vary from one laboratory to the next, and there is no co-ordination of these laboratories to establish the extent of the variability. This situation could lead to considerable differences in the measurement of aerosol concentrations and the characterisation of aerosol ensembles. Such inconsistencies are particularly important when sampling is carried out for legislative purposes to determine whether a workplace or an environmental emission complies to laid-down standards. It is also important in the quality control of some industrial processes.

In recognition of this situation, the National Measurement Infrastructure for Aerosols and Particulates in the Gas Phase was set up by the UK Department of Trade and Industry to coordinate and improve methods employed for the calibration of a range of aerosol instrumentation. This exercise constitutes part of the Valid Analytical Measurement (VAM) programme which has been charged with improving the quality and validity of all analytical measurements made in the UK.

The National Calibration Forum for Aerosol Analysis (NCFAA) forms the driving force for technical advice, supported by a number of technical projects designed to provide a range of certified test aerosol particles and recommended procedures that enable calibration processes to have some degree of traceability.

This manual is intended to provide guidance on the best methods for calibrating aerosol instruments, and represents a product of the NCFAA. Guidance is given in generic terms for eight types of aerosol instrument, with reference to specific dedicated texts for more detailed advice.

1.1 Importance of Calibration

All instruments require some form of calibration to ensure that the results they give relate to the parameter being measured and are of a consistently high quality. Ideally, all measurements should be traceable to a set of primary standards, possibly through the use of secondary standards.

1.2 Terminology

d_{ae}	Aerodynamic diameter
d_{cc}	Circumscribing circle equivalent diameter
d_{ve}	Volume equivalent diameter
d_F	Feret diameter (statistical diameter)
d_g	Geometric mean diameter
d_M	Martin's diameter (statistical diameter)
d_P	Physical diameter (geometric diameter)
d_{PA}	Projected area diameter
d_{St}	Stokes diameter
d_{DC}	Equivalent diffusion coefficient diameter
d_{EM}	Electrical mobility equivalent diameter
d_{EZ}	Electrical sensing-zone equivalent diameter
d_{LS}	Equivalent light scattering diameter
ξ	Collection efficiency
μ	Gas viscosity
σ_g	Geometric standard deviation
χ	Dynamic shape factor
NCFAA	National Calibration Forum for Aerosol Analysis
NMSPU	National Measurement System Policy Unit
SEM	Scanning electron microscopy
TEM	Transmission electron microscopy
VAM	Valid Analytical Measurement

Other terms and parameters occur only in association with specific equations, and are not included in the above list. Their definition can be more conveniently found in close proximity to the relevant equations in the main text.

1.3 Scope

Up-to-date guidance is given in the following sections on the calibration of a range of aerosol instrumentation techniques. This advice is restricted to the calibration of instruments to measure and/or select the particle size distribution of aerosol ensembles. Although aerosol concentration (number of particles m^{-3} and kg m^{-3}) represents another important ensemble characteristic that needs to be determined with confidence, calibration procedures for this parameter are not included in this document. Such valid measurements could constitute a worthwhile topic in a future manual. Furthermore, no consideration is given to the sampling and transmission efficiences of the instruments (guidance is given in an equivalent document produced during the VAM programme that covers a series of sampling guidelines).

CHAPTER 2

Measurement Techniques

2.1 Particle Size Parameters

Particle size is the most important parameter used to assist in defining the physical characteristics and behaviour of airborne particles. The size distribution of an aerosol ensemble is generally polydisperse, with sometimes up to a one hundred-fold range between the smallest and largest particles (Hinds (1982)). An appreciation of how aerosol properties can vary with particle size is fundamental to the understanding of their behaviour. It is necessary to make use of a 'particulate approach' to characterise aerosol properties in terms of the size of the individual particles which constitute the ensemble.

Aerosol particles are normally sized in terms of a characteristic dimension, or more often a diameter of a 'selectively equivalent' spherical particle. These dimensions are expressed in the units of micrometre and/or nanometre. Some authors size particles in terms of their radius rather than the more usual diameter, although the latter is to be preferred. Aerosols are often described as being in the micrometre ($\sim 1\,\mu m$) and sub-micron size range ($< 1\,\mu m$), and can vary in diameter from $0.001\,\mu m$ to greater than $100\,\mu m$. Dust particles, fungal spores and pollen are generally larger than $1\,\mu m$, and fumes and smokes are smaller.

Airborne liquid droplets are spherical (unless under hydrodynamic stress), but solid particles usually have complex geometries which makes the process of describing their behaviour more difficult. Most authors have considered only simple spherical geometries to facilitate the development of mathematical theories in describing aerosol behaviour and related phenomena. It has been necessary to make use of various 'correction factors', usually expressed in terms of a spherical 'equivalent diameter', to characterise mathematically the behaviour of non-spherical particles. In general terms, equivalent diameter is defined as the diameter of the spherical particle that has exactly the same behavioural characteristic as that of the non-spherical particle under consideration. Very often this physical property refers to a parameter which describes the aerodynamic behaviour of the particle.

2.1.1 Geometric Diameters

Microscopic examination of aerosol particles permits *direct* measurement of particle size. This procedure contrasts with *indirect* methods such as sedimentation, impaction, mobility analysis and light scattering, where the particle size is estimated from the measurement of another property which is related to size. Microscopy can be used to obtain 2-D information relating to the shape of the particle, in addition to allowing an assessment of size. The linear measurements that can be made by microscopy can be traced back to accurate optical calibrations using certified calibration graticules or grids. Microscope measurements therefore provide a fundamental basis upon which other aerosol particle measurements can be related. It is necessary to employ these alternative, indirect methods in practice, because microscopy can be rather tedious and expensive.

It is generally found necessary to assign to each particle a size based upon a 2-D projected image or silhouette. For spherical particles this is the diameter of the circular silhouette, but for geometrically complex particles it is necessary to make use of a series of 'equivalent diameters' based on the geometry of the 2-D silhouette. These 2-D based equivalent diameters are geometry dependent, and differ from the more generally applicable property-based 3-D equivalent diameters.

The smallest dimension in a 2-D image is called Martin's diameter (d_M). This parameter is the length of the line parallel to a given reference line that divides the projected area of the silhouette of the particle into two equal parts. This diameter is often referred to as a 'statistical diameter', because the value depends on the orientation of the particle, and only the mean value for all particle orientations is unique for a given particle. In practice, this orientation average is rarely estimated, and it is more common when sizing particles to measure a single d_M for each of the many particles oriented randomly with respect to the reference line.

Another statistical diameter is the Feret diameter (d_F), which is the length of the projection of the image of the particle along a given reference line, or the distance between left and right tangents that are perpendicular to this reference line.

The most commonly used equivalent diameter is the projected area diameter (d_{PA}), which is the diameter of the circle that has the same area as the projected image of the particle. This is a useful measurement because, in the 2-D sense, it is independent of the orientation of the particle. Many investigators find that the circumscribing circle equivalent diameter (d_{CC}) is also useful, representing the diameter of the circle which has a perimeter just containing the outline of the irregular particle. Once again this diameter is independent of particle orientation in a 2-D sense.

The size measurements of irregular, non-spherical particles by microscopy are frequently dependent on the ability to convert the measured 2-D geometric diameters to other equivalent diameters which better describe the behaviour of the airborne particle. Shape factors, for example, can be assigned to many geometries, and these can be used to convert geometry-based diameters into more useful sizes such as the volume equivalent diameter (d_{ve}), which is the diameter of sphere which has the same volume as the non-spherical particle; this parameter can also be associated with the diameter measured by electrical sensing-zone techniques (ranging from 0.5 to 500 μm).

Optical microscopy can be used to carry out geometric measurements in the size range from approximately 0.5 to 50 μm. Smaller sizes require the use of electron microscopy: scanning electron microscopy (SEM) for the range 0.01 to 20 μm diameter, and transmission electron microscopy (TEM) in the range 0.01 to 10 μm diameter.

2.1.2 Equivalent Diameters Based on Behavioural Properties

As opposed to the 2-D dependent geometric equivalent dimensions discussed above, the following equivalent diameters are 3-D dependent, and are related to an equivalence of a selected physical property of the non-spherical particle. The Stokes diameter (d_{St}) is one of the most important examples of the 3-D, property-based equivalent diameters, representing the diameter of a sphere that has the same density and settling velocity under gravity as the particle.

The aerodynamic diameter (d_{ae}) is of somewhat more fundamental importance in obtaining an understanding of the behaviour of airborne particles, and is defined as the diameter of the sphere of unit density which has the same settling velocity under gravitational forces as the particle.

The volume equivalent diameter (d_{ve}), the Stokes diameter (d_{St}), and the aerodynamic diameter (d_{ae}) of an aerosol particle are related in terms of particle density and shape. Under Stokesian conditions, these relationships can be expressed in terms of simple equations. All three diameters can be defined in terms of particle aerodynamic diameter, rather than particle geometry. Aerodynamic diameter is the key particle dimension for describing airborne behaviour such as dispersion, filtration, respiratory deposition, and the performance of many types of air cleaner. Instruments such as the elutriator, impinger, cyclone, centrifuge, cascade impactor and particle relaxation devices use aerodynamic separation or characterisation to measure the aerodynamic or Stokes diameter.

Electrical sensing-zone equivalent diameter (d_{EZ}) has been equated with the volume equivalent diameter (d_{ve}), although d_{EZ} is obtained by measuring the disturbance the particle makes to an electrical field developed in an electrolyte. This disturbance is dependent upon the volume of the particle, but also depends upon the electrical conductivity of the particle and the electrolyte. The method is commonly used to size powders greater than a micron in diameter.

A number of other equivalent diameters are frequently adopted to describe the size distribution of aerosols. Often the utilisation of specific techniques for carrying out measurements is ultimately dependent upon the size range of the aerosol particles under investigation. Over the 0.003 to 1 μm range, for example, the electrical mobility method is used, and the size of the particle is expressed in terms of electrical mobility equivalent diameter (d_{EM}). Particles that have been electrically charged by diffusion under well-defined conditions acquire a known charge, and a unique electrical mobility is associated with every particle size. The distribution of particle size can therefore be determined by measuring the distribution of electrical mobility.

The diffusion of an aerosol ensemble is the net transport of these particles from a region of higher concentration to a region of lower concentration. This process is controlled by the diffusion characteristics of the particle, and for particles in the size range 0.001 to 2 μm, the diffusion process can be used in laminar flow to achieve separation. The equivalent diffusion coefficient diameter (d_{DC}) can therefore be used to describe the size of small aerosol particles.

Very small particles will increase in size when placed in a supersaturated environment to become micron-sized droplets that can be used to enumerate the original nuclei. Aerosol particles in the range 0.002 to 0.2 μm diameter may be saturated and cooled by adiabatic expansion to create the conditions of supersaturation for subsequent growth. Nuclei will grow to $\sim$10 μm, regardless of their original size, and the number of droplets and hence nuclei can be determined using condensation particle counters.

Optical measurements of aerosols are extremely sensitive and nearly instantaneous, resulting in no physical contact with the particles. Light scattering by aerosol particles of $\sim$0.05 μm diameter is described by Rayleigh's theory of molecular scatter. With larger diameter particles, the light scattering process is more complex and may be described in terms of the Mie theory, in which the particle size and the wavelength of light are of the same order of magnitude.

Aerosol particles are illuminated by a beam of light which they scatter and absorb to diminish the intensity. This extinction process involves only the attenuation of the light along the projecting axis, but has been successfully adopted to define the visibility of an aerosol-containing atmosphere. Since it is impractical to analyse the refracted and reflected light scattered from particles smaller than 50 μm, the interaction of light and aerosol particles is described in terms of the angular distribution of the scattered light which is dependent on the refractive index and the size of the particle. The refractive index parameter is dependent on the wavelength of the light, and therefore this parameter should be noted when equivalent diameters are quoted using light scattering techniques.

Standard conventions have been adopted to describe the angular distribution of light scattered by an aerosol particle. Light which deviates only slightly from the incident direction has a small scattering angle and is said to be 'forward-scattering light'; light reflected or scattered back towards the source is called 'back-scattered light'; light can also be scattered anywhere between the two extremes (such as 90° to the incident light). Either type of scattering can be used, under appropriate conditions, to size single aerosol particles in terms of an equivalent light scattering diameter (d_{LS}). Optical counters are particularly effective in measuring aerosol particles in the size range 0.1 to 50 μm.

Table 2.1 *Aerosol analysers—Measurement techniques*

Particle size parameter	*Measurement technique*	*Types of instrument*	*Comments*	*Generic references*	*Example of application*
Geometric diameter	Microscopy	Optical, SEM and TEM	Choice of method dependent on size range to be measured	Bradbury (1991)	Direct measurement of 2-D geometric diameters Application of image analysis for speed
Aerodynamic diameter	Particle relaxation	APS, API, ESPART *etc.*	Dependent on particle density and shape	Wilson and Liu (1980), Marshall *et al* (1991)	Real-time particle size distributions, quality control of powders, sampler tests, monodispersity of test aerosols
Aerodynamic diameter	Inertial separation	Cascade impactors, impingers, inertial impactors, cyclone samplers, centrifuges and inertial spectrometers	Wide range of measurement by appropriate selection of instrument	Hinds (1982)	Size distribution measurements of aerosol drug delivery systems
Stokes diameter	Gravitational sedimentation	Timbrell spectrometer and elutriators/ sedimentometers	Measurements made only under viscous conditions	Timbrell (1972)	Classification of aerosols, and measurement of true d_{ae} through d_{St} and density

(*continued*)

Table 2.1 *Continued*

Particle size parameter	*Measurement technique*	*Types of instrument*	*Comments*	*Generic references*	*Example of application*
Electric sensing-zone diameter	Electrical sensing, distortion effects of particle volume on electric field	Coulter counter and ELZONE	Closely related to volume equivalent diameter but also dependent on particle conductivity	Coulter sales literature	Measure size distribution of dusts, cells, test materials
Electrical mobility diameter	Particle electrical mobility dependent on surface area	Electrical mobility analysers and differential mobility analysers	Diffusion charging required to establish relationship between particle mobility and size	Liu and Pui (1975), Knutson and Whitby (1975)	Measuring size distribution of particles $< 1\,\mu m$ diameter
Diffusion coefficient diameter	Measurement determined by diffusion coefficient of particle	Diffusion batteries and denuders	Used to separate and collect gases or vapours from airborne particulate	Knutson and Sinclair (1979), Hinds (1982)	Sub-micron particles; diffusion denuding to separate and collect gases or vapours from airborne particulate
Light scattering diameter	Single particle light scatter	Optical and laser size analysers such as the Royco and Polytec Many others	ϕ_{LS} dependent on shape and, depending on particle size, on the refractive index of the particle	Pinnick and Auvermann (1979)	Remote sensing of aerosols; fast-acting alarm systems in industrial applications

(continued)

Table 2.1 *Continued*

Particle size parameter	*Measurement technique*	*Types of instrument*	*Comments*	*Generic references*	*Example of application*
—	Laser diffraction pattern	Malvern type	Wide operational particle size range	Swithenbank *et al* (1977)	Non-intrusive measurement
Laser–Doppler related size	Laser phase–Doppler analysis	TSI and Dantec type instruments	Developed from laser–Doppler anemometry techniques: particle velocity and size	Yeoman *et al* (1982)	Non-intrusive measurement
Related to light scattering diameter	Laser intensity deconvolution analysis	Insitec PCSV type	Wide operational particle size range, useful under arduous conditions	Holve (1980)	Non-intrusive measurement
Size dependent upon light attenuation	Laser–particle interaction with image analysis	GALAI-CIS	Light attenuation sizing, independent of refractive index	Karasikov and Krauss (1988)	Useful for airborne particles and liquid suspensions

2.2 Relevance of Particle Size Parameter

2.2.1 Geometric Diameters

There are a number of microscopy-based measurements which are direct, and based on a 2-D image of the particle. They are not particularly applicable to describing aerosol particle behaviour unless they can be related to (for example) the equivalent volume diameter by the utilisation of particle shape factors.

2.2.2 Equivalent Diameters

Equivalent diameters are all based upon specific properties of the aerosol particle. The most fundamental is aerodynamic diameter (d_{ae}), which is related to the behaviour of the particle in the airborne state under laminar flow conditions, and has many aerosol-related applications.

A number of other equivalent diameters can only be related to aerosol particles within certain size ranges. Instruments that classify aerosol particles in terms of these diameters are only effective in the study of aerosol particles over particular size ranges. These instruments are many in number and type, and employ a wide range of different operating principles. Invariably, they require calibration. The relevant instrument records particle size in terms of the equivalent diameter which reflects the measuring technique being used. These relationships are listed in Table 2.1, including details of the relationship between particle size parameter (i.e. equivalent diameter), measuring technique and type of instrument.

CHAPTER 3

General Considerations

3.1 Selection of Equipment/Reference Materials

Careful consideration needs to be given to the reasons for performing any aerosol particle-size measurement. If the resulting data are not suitable for the intended application, the measurements are worthless. By way of an example, if the user is interested in the deposition of a particular aerosol ensemble in the lung, it is pointless examining the aerosol particles beneath a microscope and determining a size distribution based on an equivalent geometric diameter. The measurement may be accurate and precise, but will not be fit for purpose or relevant, as the deposition of the particles within the lung will depend upon their aerodynamic diameter and not their equivalent geometric diameter. Thus, great care must be taken in the choice of measurement technique.

Once a particular measurement technique has been deemed appropriate, it is equally important that the data produced are valid in terms of accuracy and precision. This is achieved by demonstrating traceability using appropriate reference materials. The choice of reference materials is as important as the measurement technique. All of the following points regarding equipment, traceability, calibration and reference materials need to be considered:

(a) Equipment suitability should be ascertained before any measurements are undertaken.
(b) The equipment must be capable of achieving the required accuracy levels (where such levels are specified).
(c) Equipment should only be operated by authorised and qualified staff. Written instructions and manufacturers' manuals should be readily available.
(d) Equipment should be maintained according to the manufacturers' instructions, and should be regularly re-calibrated and checked for correct function after movement from one location to another.

(e) Equipment that is damaged, defective or unfit for use should be withdrawn from service.
(f) Equipment that has been outside the direct control of a laboratory should be checked for correct function before being returned to service.
(g) Any computer system used in conjunction with an instrument should be validated before use.
(h) Subsidiary measurements (*e.g.* environmental conditions) may significantly affect the accuracy or validity of calibrations and such data need to be assessed.
(i) The procedures adopted to carry out both the calibration and the measurements must meet the requirements of BS 5781 'Measurement and Calibration Systems'.
(j) Reference standards used in a programme of calibration and measurement should be used for no other purpose.
(k) Systems of calibration and measurement should be designed and operated to ensure that all measurements are traceable to national or international standards of measurement, ensuring that confidence may be placed in the quality of measurements carried out at all levels of the traceability chain.
(l) Where the concept of traceability is not applicable in practice, satisfactory evidence of correlation within the calibration or measurement procedures should be provided.
(m) Reference standards should cover the range of measurements to be performed.
(n) An estimate of the measurement uncertainties using accepted methods of analysis should be provided for all calibrations (see Section 3.2).
(o) Use should be made of both pure reference materials and reference materials operating in matrices that match the measurement conditions.
(p) Whenever possible, reference materials should be used that have been produced and characterised in a technically valid manner. Certification should provide evidence of traceability to national or international standard reference materials.
(q) When a certified reference material is not available, reference materials with suitable properties and stability should be used.
(r) If standards are prepared from materials of known properties or uncertified standards are purchased, the user must verify that the standards are of acceptable quality and suitable for the purpose.

3.2 Uncertainty of Measurement

Every measurement has an associated uncertainty, which must be quantified for the measurement to have any validity. The following points should be considered in determining this uncertainty:

(a) The uncertainty should include an estimate of the uncertainties identified with the methods and procedures.

(b) All significant uncertainties in the measurement process should be taken in account, including the uncertainties attributable to the measuring equipment, reference measurement standards (including material used as a reference standard), staff using the equipment, measurement procedures, sampling and environmental conditions.
(c) Data obtained from internal quality control schemes and other relevant sources should be included.
(d) The limits chosen for the calibration of measuring equipment must cover the conditions under which the equipment is to be used.
(e) When estimating the uncertainty of the calibration process, the cumulative effect of each successive stage of the calibration procedure must be assessed.

3.3 Traceability and Documentation

Appropriate and traceable calibration is vital to ensure the validity of any subsequent measurements. A valid calibration programme has to be applied to the primary measuring equipment and any subsidiary measurements that may affect the accuracy or validity of the particle size analysis. The aspects listed below have to be addressed:

(a) Appropriate methods and procedures should be used for all calibrations; they should be consistent with the accuracy required and with any standard specifications relevant to the calibration concerned.
(b) Wherever possible, methods and procedures that are established and up-to-date should be used. When other methods or procedures are used, it should be demonstrated that they are suitable for purpose and should be clearly documented.
(c) Laboratories should have access to and maintain on record the information necessary for the proper performance of calibrations.
(d) Methods and procedures for calibrations, instructions for the operation and calibration of equipment, and any information under 3.3(c) above should be documented to ensure proper implementation and consistency of application from one occasion to another. Such records should be readily accessible.
(e) The integrity of the calibration data should be guaranteed. Procedures should be established to ensure that the collection, entry, processing, storage and transmission of calibration data are in accord with the guidelines given above.
(f) Calculations and data transfers should be subject to appropriate checks.
(g) When using the services of an external organisation to calibrate equipment, laboratory staff should satisfy themselves that the organisation in question has satisfactorily addressed the issues raised in this Section.

3.4 Calibration Intervals

The choice of the time interval between calibration campaigns is a matter for each individual laboratory. Calibrations should be performed at a frequency which ensures the validity of the data obtained, but not so frequently that they become too severe a burden on time and resources. Staff should consider:

(a) requirements of any relevant standard specifications for the measurements involved,
(b) recommendations of the equipment manufacturer,
(c) type and stability of the equipment,
(d) extent and regularity of equipment use,
(e) influence of environmental conditions (*e.g.* temperature, humidity and vibration),
(f) desired accuracy of the measurements for the application concerned,
(g) trends determined by examination of the records of previous calibrations,
(h) evidence for calibration requirements obtained from service and maintenance records,
(i) any known or observed tendency for the instrument to exhibit wear or to drift during use,
(j) information obtained from in-house checks, using known standards.

The following criteria also need to be met:

(k) Calibrations should be sufficiently frequent to minimise the risk that the results of any measurement performed between calibrations may be affected because some of the equipment has failed to perform to the specified requirements and the problem has not been rectified. For certain types of measurement, calibration is necessary as part of normal operations using appropriate certified reference materials.
(l) When new items of equipment are used with only limited information available, the calibration interval selected initially should be shorter than eventually expected. The interval may then be adjusted at a later date as greater assurance is established from the information obtained from the initial calibrations.
(m) Maintenance and calibration timetables should be periodically reviewed to take into account the variation in type, conditions and frequency of use of the measuring equipment.
(n) When the performance of the measuring equipment deviates from the specified requirements, the maintenance and calibration intervals should be reviewed immediately and modified where necessary. Such equipment should not be returned to service until the reason for the deviation has been established and eliminated prior to recalibration.
(o) The time intervals between calibrations (and maintenance where appropriate) should be shortened when the results of preceding calibrations or

intermediate checks indicate that the equipment is no longer performing in accordance with the specified requirements.

(p) The time intervals between calibrations should only be increased when the results of preceding calibrations and any intermediate checks show that the performance of the equipment is likely to remain within the specified requirements throughout the new time period.

3.5 Records

Appropriate records are an essential part of any calibrated system. All time and resources invested in calibrations are negated unless sufficient documentary evidence is available to prove the validity of the calibrations. Suitable written records should address the following issues:

(a) Records should be kept for each item of measuring equipment, including details of the reference measurement standards, reference material standards and test equipment used in the calibration exercises. This information should include evidence that the calibration is traceable, either through in-house documentation or calibration certificates from external organisations.

(b) All records should contain detailed information of the equipment/reference material used for calibrations, and also a full and up-to-date history of the calibration of this equipment/reference material.

(c) Sufficient information should be available to demonstrate the measurement capability and traceability of each item of measuring equipment, and the range of each reference material, shelf-life and storage conditions.

(d) Each record should include or refer to:
- (i) calibration method or procedure adopted, and the relevant standards used,
- (ii) date on which each calibration was performed,
- (iii) calibration results obtained after and, where relevant, before any adjustment and repair,
- (iv) specified calibration interval,
- (v) limits of permissible error,
- (vi) calibration certificates,
- (vii) documentation for all reference materials used for calibration, providing evidence of characterisation of the material, and the traceability to national or international standards of measurement or to national or international standard reference materials,
- (viii) environmental conditions at the time of the calibration, and the corrections made for such conditions, if necessary,
- (ix) statement of the uncertainties of measurement involved in the calibration and of their cumulative effect,
- (x) any design or performance specifications,

(xi) names of persons who perform the calibration and check the results,
(xii) any limitations in the calibration data obtained,
(xiii) details of any maintenance, servicing, repair or modifications carried out, particularly at the time of calibration.

(e) Similar records should be maintained of any checks carried out on equipment or reference materials between calibrations.

CHAPTER 4

Ancillary Considerations

Table 4.1 lists the steps in a calibration procedure. The various stages trace the path from the generation of the test aerosol through air flow calibration, measurement and data handling to instrument calibration. Both the operating principles and limitations of all the various types of aerosol analysers are not described in this manual, but descriptions of the more important of the analysers provide a useful guide to the current method(s) of calibration. More comprehensive review articles on the subject have been written by Liu (1976), Fuchs ((1978) impactors only), and Fissan and Helsper (1982).

The two parameters of greatest importance in the calibration of aerosol analysers are particle concentration and size distribution. However, reference is normally made to only the calibration of the particle size axis of an aerosol analyser, although any analysis requires quantification of the particle number or mass/volume related parameters. For instance, inertial devices such as impactors and cyclones are generally used to determine the mass concentration and size distribution of aerosols, since the deposition pattern is more easily assessed by gravimetric-based analytical methods than by microscopic examination of individual particles. Some types of optical particle counters monitor individual particles to derive number concentrations and size distributions. Number-size distributions can be converted to mass-size distributions and *vice-versa*, but large errors and uncertainties can occur in such manipulations, particularly at the extremes of the size distributions. Therefore, it is good practice to choose a calibration method which is appropriate to the type of data expected from the aerosol analyser. Mass-size calibrations require larger quantities of aerosol than number-size calibrations, unless a sensitive tracer technique is used such as the addition of a fluorescent dye to the aerosol particles. Calibrations based on mass analysis are more sensitive to fluctuations in the number of particles larger than the mean size; calibrations based on particle counting are more sensitive to variations in the number of particles smaller than the mean. Therefore, the removal of outlying particles that are significantly larger or smaller than the

Table 4.1 *Generic calibration procedure*

Stage	*General considerations*	*References*	*Comments*
Generation	Traceability of CRMs	Standards — ISO, CEN	Monodisperse CRMs — size range, quantity and generator
Flow calibration	Traceability of flow measurement	Standards — ISO, CEN, BSI, scientific publications	Bubble flowmeter, wet gas meter and critical orifices
Measurement	Check manufacturers' instructions		Measurement strategy — frequency, size range, *etc.*
Data handling	Estimate of errors, and statistical analysis		Define whether mass- or number-based; complete curve or d_{50} only
Calibration	Initial calibration or periodic checks		

mean size may be a necessary requirement in the preparation of a calibration aerosol.

The validity of particle size and concentration measurements made with aerosol analysers depends upon the instrument being operated correctly. Although particle behaviour in many aerosol samplers can be described quite accurately from first principles (*e.g.* impactors), it is frequently necessary to check the operational characteristics of such instruments with well-defined particles in order to verify that they are working according to theory. This specification is especially true when the measurements have to meet quality assurance standards (*e.g.* BS EN ISO 9001 : 1994) or conform to legal requirements.

Aerosol analysers need to be calibrated for four reasons:

(a) to ensure that the instrument is functioning correctly (routine quality assurance),
(b) to compare performance with theoretical predictions,
(c) to compare performance with other analysers of the same type,
(d) to compare measurements of a given aerosol with other instruments working on different principles.

The final comparison is often of major importance when undertaking several different measurements of an unknown aerosol, since the geometric diameters of many particles may be quite different to their aerodynamic diameters or the equivalent size parameters obtained by light scattering or electrical mobility measurements.

All of the transport properties of aerosols are strongly dependent upon particle size, and the establishment of these relationships is the main aim of many studies involving aerosol analysers. Most environmental aerosols are polydisperse, and it is necessary to assume that the properties of a polydisperse aerosol with a mean particle diameter coincide with the corresponding properties of a monodisperse aerosol with the same diameter. Although there has been a tendency to use aerosols with high monodispersity for calibrations, it has been recognised that calibration using a well-characterised polydisperse aerosol can be a valuable exercise, particularly if the sensitivity of the instrument is affected by a non size-dependent property of the aerosol such as refractive index.

The particle size distributions of many aerosols approach log-normal, and the degree of dispersity can be conveniently characterised by the standard deviation (σ_g):

$$\ln(\sigma_g) = \left[\frac{\sum_{i}^{N} \ln\left[\frac{d_i}{\bar{d}_g}\right]^2}{N} \right]^{\frac{1}{2}}$$

where N is the number of particles in the sample each with diameter d_i, and $\bar{d}_g$ is the geometric mean diameter given by:

$$\bar{d}_g = \left[\prod_{i}^{N} (d_i) \right]^{\frac{1}{N}}$$

For most practical purposes an aerosol is considered 'monodisperse' if σ_g is less than 1.2, and many aerosol generators are capable of producing aerosols with σ_g values smaller than 1.1 within the optimum capabilities of their size range.

4.1 Aerosol Generation

Most aerosol studies are carried out with particles ranging in size from about 0.01 to 100 μm diameter, and no single aerosol generation technique can produce particles spanning the entire range (examples are given in Figure 1). Factors such as aerosol concentration and the morphology of solid particles must also be considered in the choice of a suitable aerosol source. This section constitutes a brief summary of the main features of the more commonly used aerosol generators, and is intended as a guide in the choice of method used to produce calibration aerosols.

A few general comments can be made before describing the individual

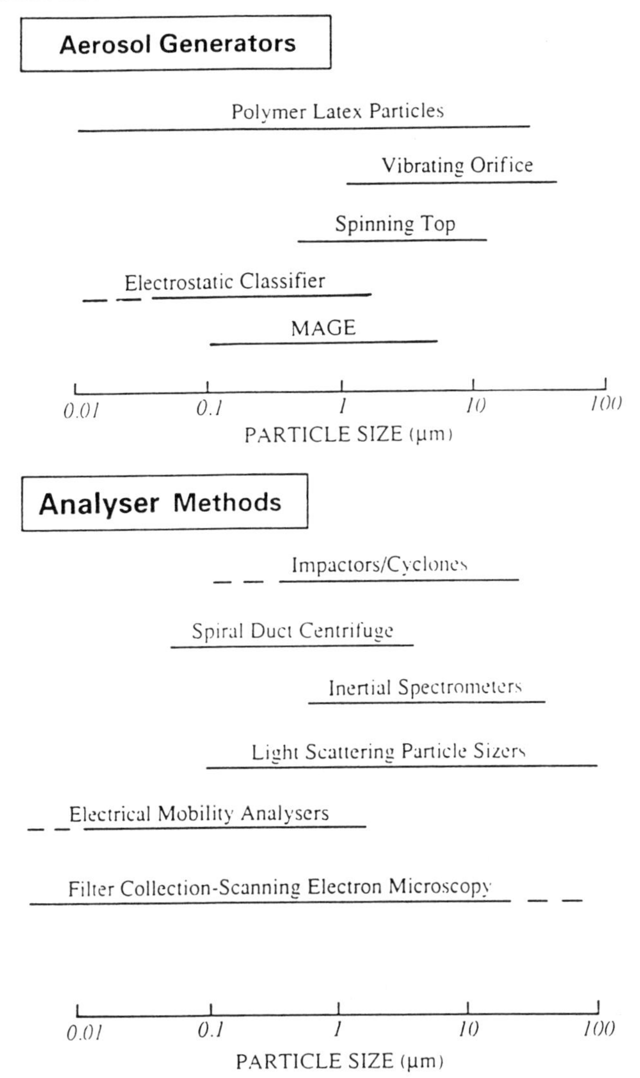

Figure 1 *Size ranges of common calibration aerosol generators compared with various aerosol analysers*

techniques. Polydisperse aerosols are normally produced by direct dispersion from a dust generator, of which there are many types, as well as by liquid atomisation. Monodisperse particles that have d_{PA} values larger than 0.5 μm can be produced by controlled atomisation of a liquid or by resuspension of preformed polymer latex particles. Monodisperse aerosols can also be produced by classification of a polydisperse aerosol on the basis of particle inertia or electrical mobility. Condensation aerosol generators of the Sinclair-LaMer or MAGE type have a useful size range from 0.1 to 5 μm diameter, while methods based on classification by differences in electrical mobility are only suitable for producing sub-micron particles.

There are a number of comprehensive reviews of aerosol generation for calibration purposes (*e.g.* Fuchs and Sutugin (1966), Dennis (1976), Raabe (1976), Willeke (1980), and Liu (1976)). Dennis (1976) has written a general review of the subject, while the paper by Fuchs and Sutugin (1966) concentrates on condensation aerosol generators, and the review by Liu (1976) is concerned with aerosol generators developed at the University of Minnesota. Raabe (1976) has included information on the correct use of polymer latex particles. Although the book edited by Willeke (1980) is mainly concerned with the generation of aerosols for animal exposure facilities, there is much useful information on monodisperse aerosol generation techniques.

4.1.1 Monodisperse Aerosols

Polymer Latex Particles

Polymer latex aerosol generation is the simplest technique, since the particles are pre-formed and need only to be suspended in air or other gaseous media. Monodisperse microspheres of polystyrene and related polymers ranging in diameter from 0.038 to 20 μm are available from several commercial sources including Seragen Inc., Indianapolis, USA; Duke Scientific (Europe), Hilversum, Netherlands (through Brookhaven Instruments Ltd, Stock Wood, Worcs. in the UK); Polysciences Inc., Warrington, PA, USA (through Park Scientific Ltd, Moulton Park, Northampton in the UK); JSR, Japan; and Dyno Industrier a.s., Oslo, Norway. These particles can be obtained in certain sizes traceable to national standards of length, and are normally supplied as an aqueous suspension containing between 0.1 and 10% *w*/*v* polymer together with a water-soluble inorganic emulsifying agent and an inorganic surfactant stabiliser (as much as 6% of the weight of polymer). After diluting the original suspension, the particles are suspended as an aerosol using a gas-driven nebuliser (Raabe (1976)). The dilution required to produce one polymer latex particle per water droplet depends chiefly on the size of the particles and to a lesser extent on the water droplet size distribution (Raabe (1968 and 1976)). Thus to produce a singlet to multiplet ratio of 0.95, 2 μm diameter latex particles must be diluted from the original 10% *w*/*v* stock by a factor of 15, assuming the water droplet size distribution has a volume median diameter (VMD) of 5 μm and σ_g of 1.2. The dilution factor increases to 80 for the same VMD with a σ_g of 2.0. However, 0.2 μm diameter latex particles must be diluted more than 15 000 times, assuming the same original water droplet size distribution. Although the formation of multiplet particles is undesirable for many calibrations, they can be used to provide additional information about the performance of many inertial devices, since the physical and aerodynamic size relationships between singlet and multiplet aggregates containing up to eight particles are well known (Stöber and Flaschbart (1969)).

Aerosol analysers can be calibrated more quickly if a well-defined mixture or cocktail of polymer latex microspheres can be used. Such cocktail reference materials consist of a blend of different sizes of monodisperse particles in which the relative amounts of the components are known. User needs for these materials have been evaluated as part of a market survey for the VAM programme (Lewis *et al* (1993)). The most popular choices for cocktail reference materials were found to be three-component mixtures covering the size range from 0.1 to 10 μm volume equivalent diameter, which were requested to be made available in 20 ml vials (containing 2.5% *w/v* polystyrene latex). Four such cocktails were prepared as part of the VAM project (Marshall (1995a)), and details of their specifications and availability are given in Appendix A.

Water of the highest purity should always be used for the dilution of suspensions since dissolved impurities will form separate sub-micron particles, as well as coating the surfaces of the latex particles to make them larger (Reist and Burgess (1967), and Langer and Lieberman (1960)). This problem is difficult to overcome because the latex particles agglomerate irreversibly if the stabiliser is removed. One possibility is to centrifuge the latex particles from the original suspension before re-suspending them in pure water. Fuchs (1973) has summarised the precautions to be taken when using polymer latex particles: in practice, impurities do not cause a significant problem when working with particles larger than 0.5 μm diameter.

Many different designs of polymer latex aerosol generator have been formulated, and a 'home-made' example is shown in Figure 2. Aqueous latex suspension is sprayed into a suitably designed mixing chamber using a pneumatic nebuliser, and operating at an air pressure in the range 105 to 350 kPa (15 to 50 psig). The choice of nebuliser can be important, particularly for studies where constant aerosol concentration must be maintained for more than a few minutes, (*e.g.* a Retec X-70/N nebuliser). However, the liquid reservoir in the Retec is not agitated when in use and particles sediment from suspension, resulting in a decrease in the generated aerosol concentration with time (*e.g.* reduction of 10% in mass concentration in 15 min is typical for 2.8 μm diameter particles). The liquid is continuously agitated in a DeVilbiss type-40 nebuliser, and therefore this device is likely to be more suitable for work in which the stability of the aerosol concentration is important. After leaving the mixing chamber, the polymer latex particles pass downwards to the collection point through an electrostatic charge equilibrator (line source containing radioactive krypton-85), which equilibrates (and reduces) to a Boltzmann distribution the high electrical charge imparted to the liquid droplets when they are atomised. The particles are dried by passing them through a short heated section (at about 100 °C), aligned vertically to transport particles larger than 5 μm diameter with high efficiency. Other designs of polymer latex aerosol generator have a horizontal particle-transport arrangement, which often includes a diffusion drier filled with a suitable desiccant (such as silica gel) to dry the particles.

Polymer latex particles have a well-defined refractive index when made from polystyrene, polyvinyltoluene and styrenedivinylbenzene (Garvey and Pinnick (1983)), and they have become recognised as standards for the calibration of

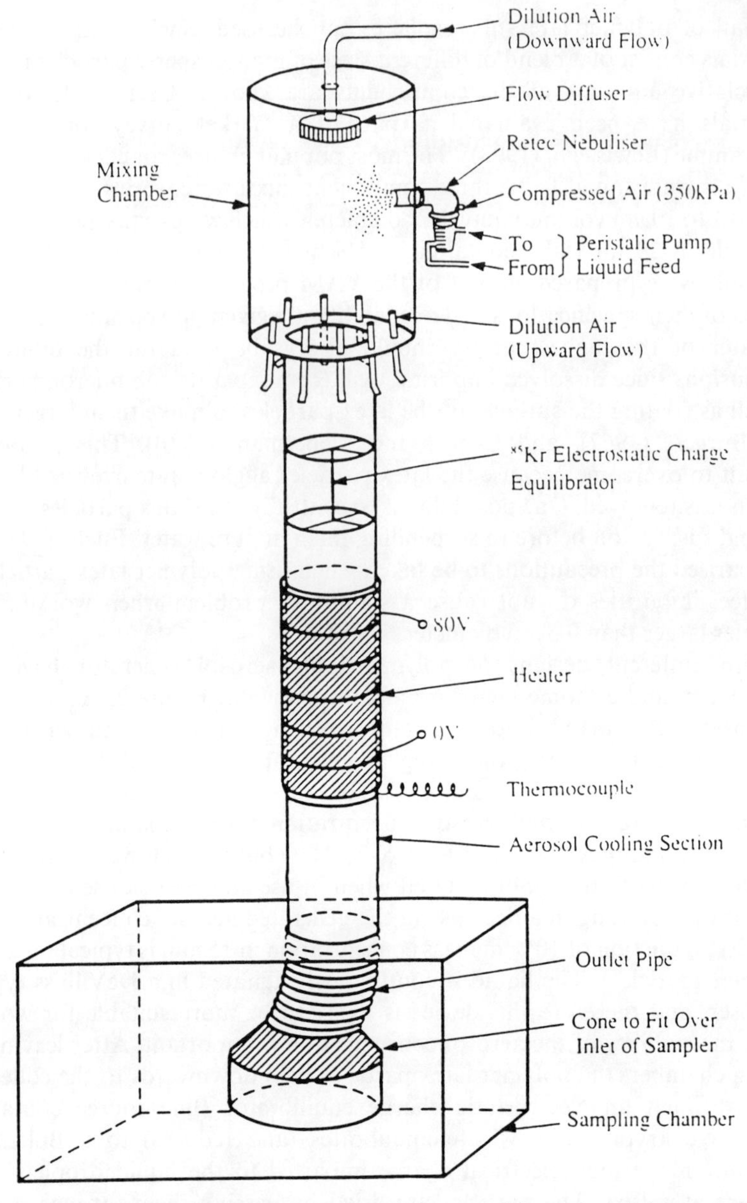

Figure 2 *Polymer latex aerosol generator with downward dispersion of the aerosol*

optical particle size analysers (*e.g.* BS 3406 : Part 7). These particles are virtually non-porous, with densities of 1.027 and 1.050 g cm^{-3} for polyvinyltoluene and polystyrene respectively, and their aerodynamic diameter can therefore be determined directly from measurements of the physical size by microscopy.

Vibrating Orifice and Related Aerosol Generators

When a thin jet of liquid is emitted from an orifice under pressure, this stream is unstable and will disintegrate into discrete droplets by the action of external forces (gravity and surface tension), as described by Rayleigh (1878). The collapse of such an unstable stream into very uniform droplets is easily attained by the application of a periodic vibration of suitable amplitude and frequency. Several aerosol generators which work on this principle have been developed in recent years, including the Fulwyler droplet generator (Fulwyler *et al* (1973)) in which the oscillator is coupled to the orifice through the liquid reservoir, an aerosol generator described by Hendricks and Babil (1972) in which the droplets are produced by disruption of a liquid jet leaving a vibrating capillary tube, and the vibrating orifice aerosol generators (VOAG) described by Ström (1969) and Berglund and Liu (1973). The last method has become the most popular technique of this type, and is commercially available (Thermosystems Inc. (TSI), St Paul, Minnesota, USA (see Figure 3)).

The VOAG can generate highly monodisperse aerosols consisting of solid or liquid particles with σ_g values less than 1.05, by feeding either a solution of known concentration or a low-volatile liquid through an orifice plate (typically 5 to 50 μm diameter). This orifice plate is mounted inside a piezoceramic crystal so that when an a.c. signal is applied, the crystal vibrates in the vertical plane

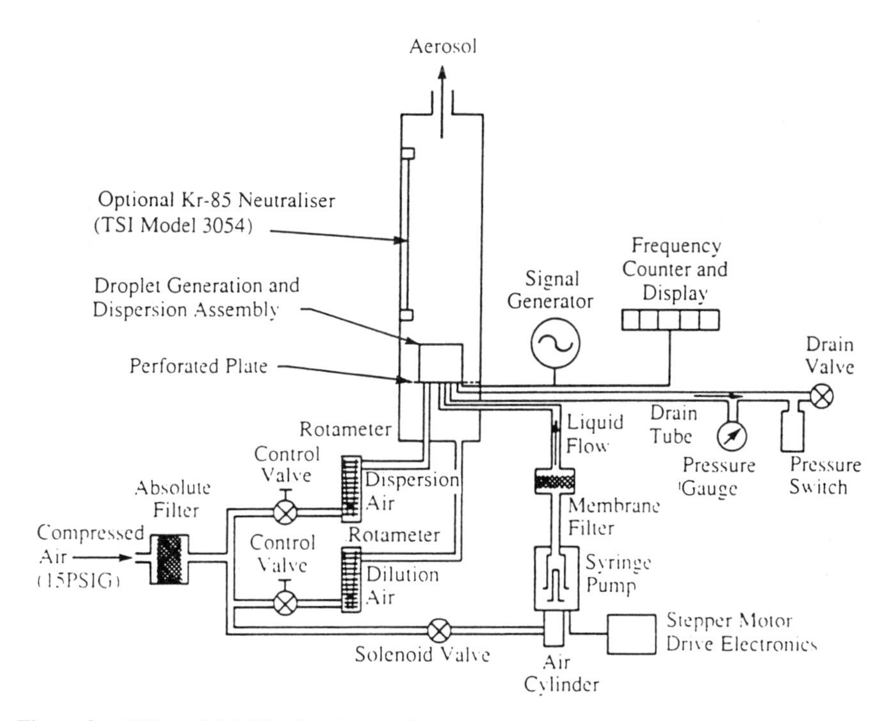

Figure 3 *TSI model 3450 vibrating orifice aerosol generator*

and disrupts the emerging liquid jet. The resulting droplets are immediately dispersed in a small flow of air (typically 0.75 to 1.5 l min^{-1}), and the desired calibrant aerosol is produced after evaporating any solvent from the liquid aerosol in a flow of dilution air. The diameter (d_P in μm) of the aerosol particles can be predicted directly from the operating parameters of the VOAG in accordance with the equation:

$$d_P = \left[\frac{10^{11} QC}{\pi F}\right]^{\frac{1}{3}}$$

where Q is the liquid feed rate (cm^3 min^{-1}), C is the fractional concentration of solute in the feed liquid, and F is the crystal vibration frequency (Hz). Optimum operating conditions for frequency and liquid flow rate as a function of orifice size have been determined by Wedding and Stukel (1974) and Wedding (1975) to predict the limits for the generation of monodisperse particles.

Although the VOAG is easy to operate, several precautions need to be taken when the device is used to produce solid calibrant particles:

(a) Liquid pressure behind the orifice must be carefully controlled.
(b) The purest reagents must be used, particularly when producing small particles from dilute feed solutions.
(c) Droplets must be carefully dried to guarantee the production of solid spheres.
(d) The resulting particles may be porous, particularly if the drying conditions are not well controlled, and this will affect their aerodynamic properties.

A VOAG system has been developed during the VAM studies to produce calibrant aerosols at a constant mass and number concentration (Booker and Horton (1995)). The system is shown in Figure 4, and further details of this work can be found in the relevant report.

The bulk density of the particles (ρ_p) can be obtained by comparing the aerodynamic (d_{ae}) and physical (d_P) diameters of spherical particles whose dynamic shape factor (χ) is unity, since:

$$\rho_P = \left[\frac{d_{ae}}{d_P}\right]^2 \left[\frac{C(d_{ae})}{C(d_P)}\right] \rho_0 \chi$$

where ρ_0 is unit density, and $C(d_{ae})$ and $C(d_P)$ are the Cunningham slip correction factors which are close to unity for particles larger than 2 μm aerodynamic diameter. The aerodynamic diameters of methylene blue particles have been measured using an Aerodynamic Particle Sizer (APS33B, Thermo-systems Inc., St Paul, Minnesota) and compared with equivalent diameters obtained by microscopy. These measurements showed that the particles were slightly porous, with values close to 1.10 g cm^{-3} compared with the bulk density

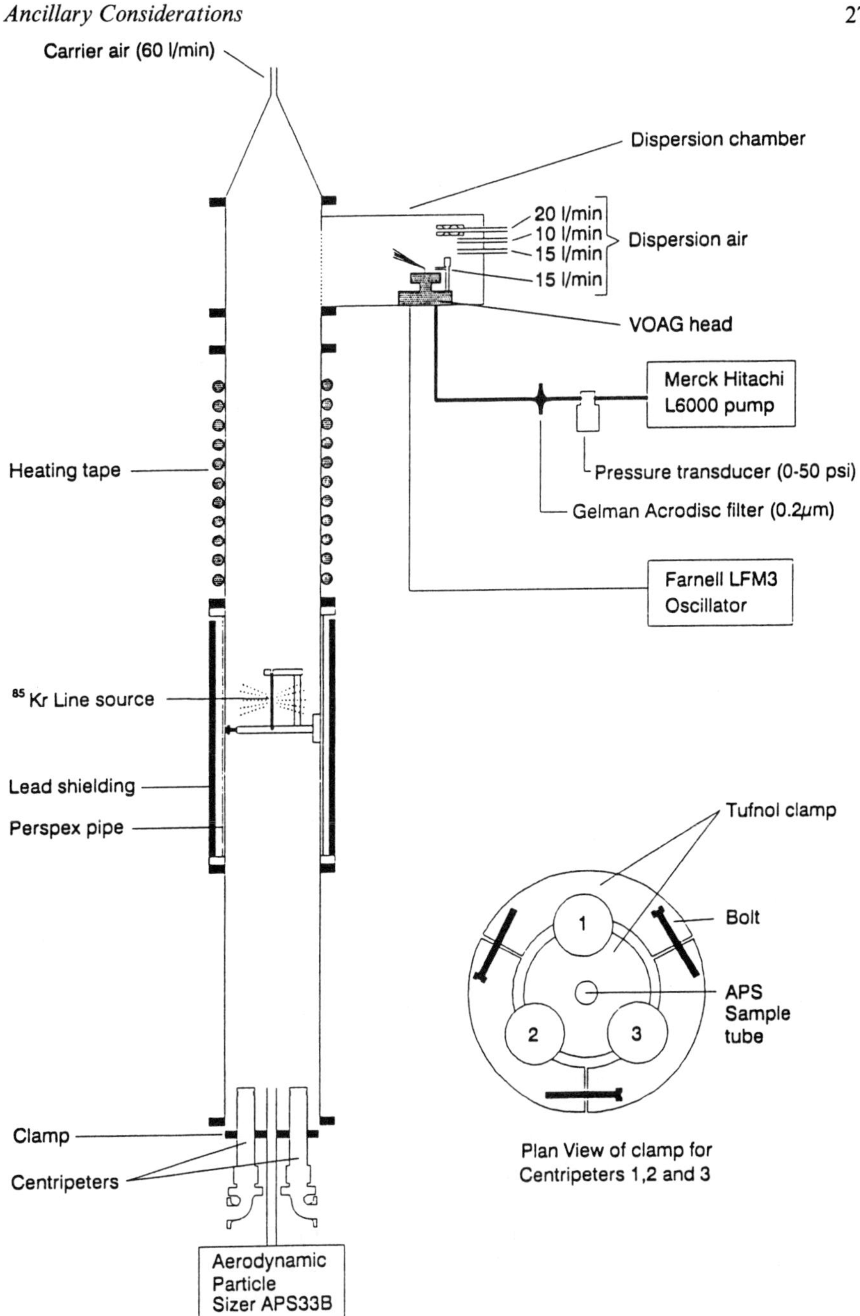

Figure 4 *Number concentration test facility*

of 1.26 g cm^{-3} quoted by Hinds (1982). A particle density of 1.10 g cm^{-3} was also obtained by O'Connor (1973) for methylene blue particles prepared by a spinning top aerosol generator. Chen and Crow (1986) have produced highly monodisperse ($\sigma_g < 1.02$), non-porous ammonium fluorescein microspheres with d_P values between 3 and 17 μm using an inverted VOAG, and working with low flow rates for the dilution air (15 to 30 l min^{-1}).

The morphology and overall shapes of solid particles produced by the evaporation of solvent from the initial droplets are determined by the presence of nuclei around which solid formation is initiated, the crystallinity of the solid being formed, and the solvent evaporation rate. Leong (1981) proposed that the optimum conditions for solid sphere formation should be a nuclei-free environment, a high solubility solvent and low drying rates. The optimum drying conditions vary for different particle types and solvents, and some exploratory studies are always required when using this technique. For instance, Vanderpool *et al* (1984) have extended the upper size limit to 70 μm diameter for ammonium fluorescein spheres by drying the particles slowly in humidified air. However, when a VOAG was used to prepare sodium chloride aerosols from aqueous solutions, the resulting particles were quasi-spherical aggregates composed of several smaller cubic crystals (Berglund and Liu (1973)).

The VOAG can also be used to generate monodisperse liquid droplets and many workers have fed low-volatility liquids such as DOP or di(2-ethylhexyl) sebacate (DEHS) dissolved in a suitable solvent through the vibrating orifice. Liquid aerosol containing a trace of fluorescent dye (McFarland *et al* (1977)) is a popular method for calibrating cascade impactors when particle bounce can be a problem. The droplets are unaffected by the small concentration of dye incorporated in the feed solution and only small amounts of aerosol need to be collected for the calibration, since the distribution of the droplets within the impactor can be accurately obtained by highly-sensitive fluorometric analysis.

Spinning Top/Disc Aerosol Generators

Monodisperse droplets can be generated when centrifugal forces disrupt a film of liquid as it spreads from the centre of a horizontal disc rotating about a vertical axis. Several versions of this type of aerosol generator have been constructed which fall into two main classes: spinning disc generators, in which the rotor is driven mechanically, and spinning top aerosol generators, in which the rotor is driven by compressed air jets. Good examples of the former type have been developed by Whitby *et al* (1965) and Lippmann and Albert (1968), while the latter method is represented by the May spinning top shown in Figure 5 (May (1949 and 1966)). Both types of aerosol generator are alternatives to the VOAG, and have the advantage that a higher output of particles is possible. Thus, the liquid feed rate of the May spinning top aerosol generator (STAG) can be increased to about 1 cm^3 min^{-1} without affecting the monodispersity of the resulting aerosol significantly, although feed rates between 0.2 and 0.8 cm^3 min^{-1} are more usual (Stahlhofen *et al* (1979), Maguire *et al* (1973) and Hurford (1981)). A further advantage of the STAG is that it can be operated continuously for several hours if the rotor surface does not become heavily coated with deposits from the feed solution, which can affect the degree of wetting of the rotor by the liquid. In contrast, the orifice of the VOAG

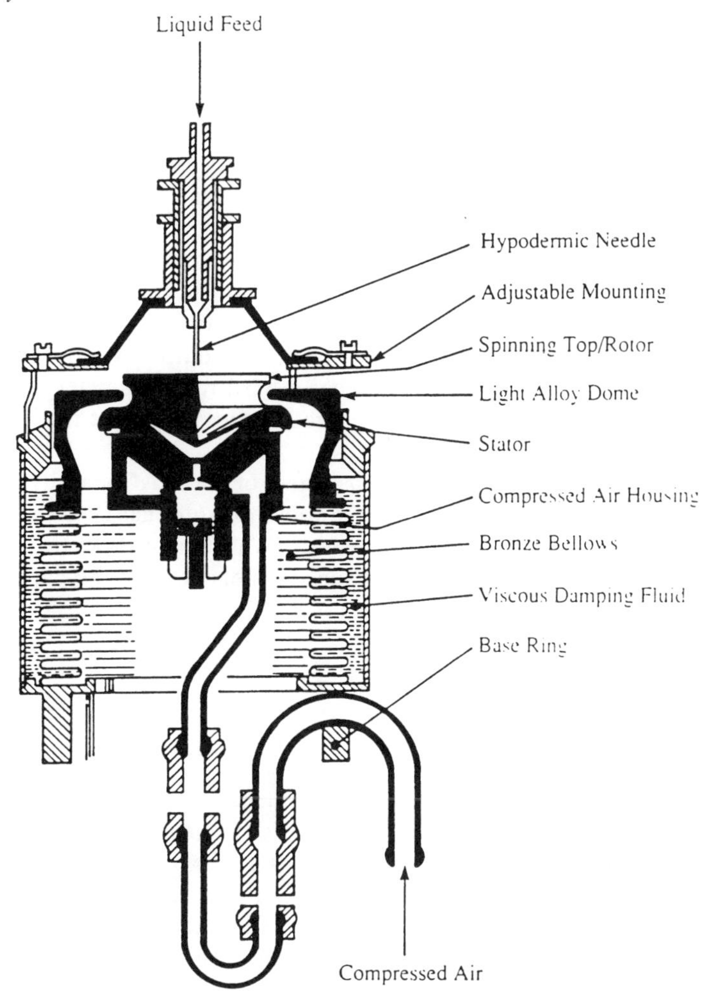

Figure 5 *May spinning top homogeneous spray*
(Reprinted by permission of K. R. May and IOP Publishing Ltd. from 'Spinning-Top Homogeneous Aerosol Generator with Shockproof Mounting', by K. R. May, *Journal of Scientific Instruments* **43**, 841. Copyright 1966 by IOP Publishing Ltd.)

requires frequent cleaning since it is easily blocked, particularly when concentrated feed solutions are used.

Spinning top and disc aerosol generators produce smaller satellites as well as the monodisperse primary droplets. These satellites form as the thread of liquid is drawn out behind each primary droplet and collapses (Walton and Prewett (1949)). Several satellites are produced with every primary droplet and, as they are on average about a quarter of the size of the latter, they can be easily removed by inertial separation soon after their formation. This is achieved by withdrawing the air close to the rotor through a separate exhaust in the generators designed by Whitby *et al* (1965). The primary droplets have greater inertia and are able to escape this suction to be transported by a flow of carrier air to the main sampling point.

While the remainder of this section is concerned with the STAG (Research Engineers Ltd, London), almost all of the operating principles apply to spinning disc type generators. Aerosols that have the same dispersity, particle concentration and size ranges can be made with both systems. A final point is worth noting with respect to the spinning disc generator: the rotor diameter can be made as large as desired (within the limit of the mechanical strength of the rotor material), making it possible to increase the aerosol output without sacrificing monodispersity; in contrast, the STAG has a fixed rotor diameter of 2.54 cm, and significant changes in the aerosol characteristics cannot be made without rebuilding the complete assembly.

The feed liquid is delivered to the rotor surface of the STAG via a hypodermic needle and careful alignment of this needle above the rotor has been shown to be critical for correct operation of the generator (Walton and Prewett (1949) and Mitchell (1984)). The equation relating the mean diameter of the particles ($\bar{d}_P$) to the operating conditions is:

$$\bar{d}_P = \left[\frac{K}{2\pi F}\right]\left[\frac{T}{D\rho_l}\right]^{\frac{1}{2}}\left[\frac{C}{\rho_s}\right]^{\frac{1}{3}}$$

where K is a numerical constant, F is the frequency of revolution of the rotor (Hz), T is the surface tension (dyn cm^{-1}), ρ_l is the density of the feed liquid (g cm^{-3}), D is the diameter of the rotor (cm), and ρ_s is the density of the solute or suspended solid in the feed liquid (g cm^{-3}). The sizes of particles produced cannot be predicted simply from the equation unless the value of K is known for the system. K is a complex factor which depends on a number of ill-defined parameters such as the roughness of the rotor and type of material being atomised. Such effects cannot easily be quantified, and K can vary considerably from one generator to another.

Cheah and Davies (1984) reported severe reductions in the aerosol output of the STAG when the primary droplet diameter was smaller than about 18 μm. For many organic liquids that have a low surface tension, the primary droplet diameter at normal rotational frequencies (*ca.* 1000 rev s^{-1}) is often smaller than this size. Cheah and Davies used streak photography to show that most of these droplets were lost when they become entrained by the in-flow of air created as the spent rotor-drive air escaped. They modified their STAG so that the radius of the stator support assembly was slightly reduced, and the height of the cover above the stator could be adjusted to decrease the amount of suction without lowering the cover beneath the trajectories of the outward moving primary droplets. These modifications resulted in the production of primary droplets in the size range 5.2 to 18 μm diameter from DEHS solutions in hexan-1-ol, with yields several times greater than had previously been achieved. Yields exceeding 70% of the theoretical maximum were obtained when generating droplets between 8 and 18 μm diameter.

Several other useful modifications have been reported in the literature. For instance, the STAG system developed by Bailey and Strong (1980) included a centripeter to concentrate the aerosol particles before use. Philipson (1973) and Garland *et al* (1982) suspended their spinning discs/tops in vertical columns to improve the particle transport. Yields in excess of 50% have been achieved with a system used to generate both ferric oxide and cerium oxide microspheres with d_P values ranging from 1 to 7.5 μm (Jenkins *et al* (1987)). Particles were calcined in a tube furnace at 700 °C using the layout depicted in Figure 6, to give the quality of microspheres shown in Figure 7. Factors such as range of particle

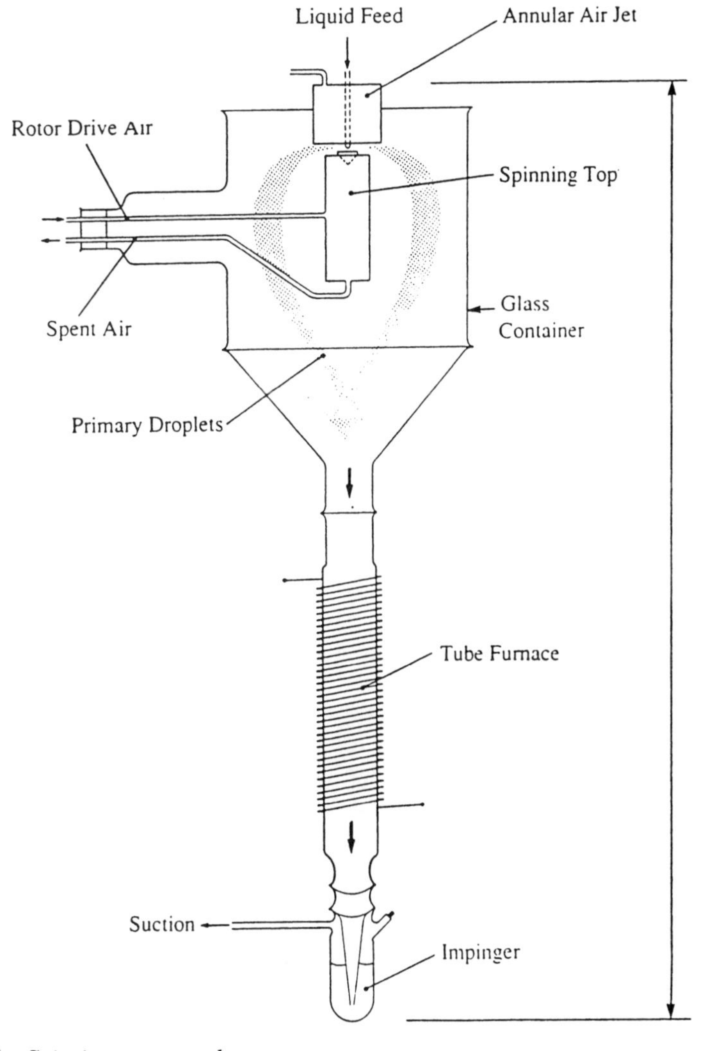

Figure 6 *Spinning top aerosol generator*

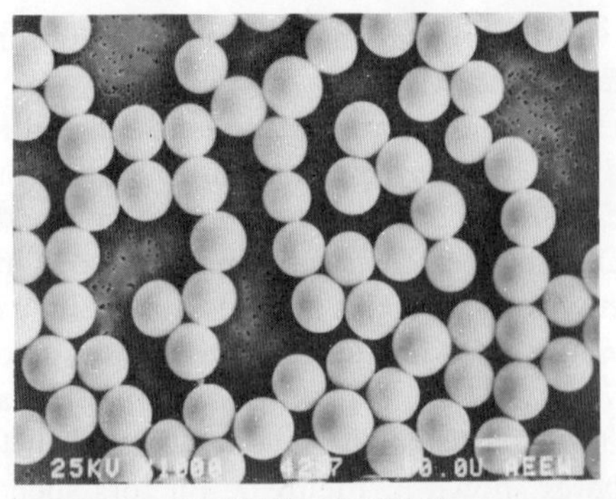

(a) 11·2µm PARTICLES; σ_g OF 1·07

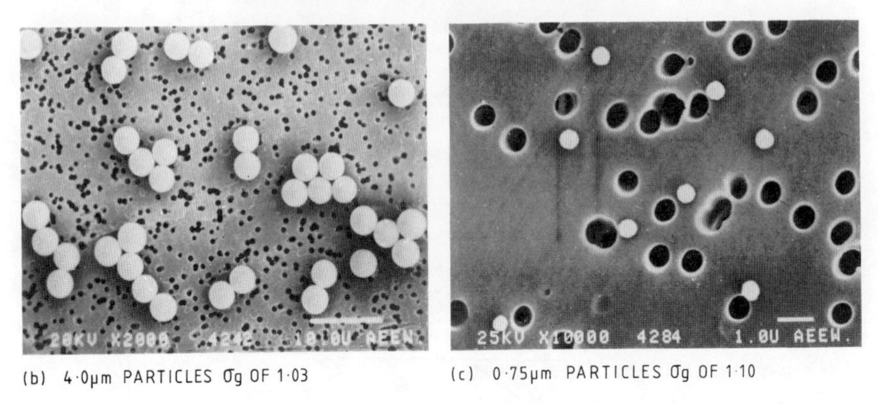

Figure 7 *Ferric oxide particles produced using the spinning top aerosol generator*

size, concentration, and the immediate or delayed use of the particles will influence the choice of generator design. Further information on this popular technique can be found in the reviews by Fuchs and Sutugin (1966) and Davies and Cheah (1984).

Condensation Aerosol Generators

Sinclair and LaMer (1949) and Rapaport and Weinstock (1955) have described aerosol generators that are frequently used to produce monodisperse particles by condensation of a vapour on foreign nuclei. The operational principles of both aerosol generators are defined in detail by Fuchs and Sutugin (1966) and Dennis (1976), and only the main features are discussed below. The Sinclair–LaMer generator constructed by Muir (1965) involves the passage of a stream of oxygen-free nitrogen through a 2 l flask in which a high boiling-point liquid such as DEHS is heated and stirred continuously. The temperature is held constant to within ± 0.5 °C as monitored by a contact thermometer. Nuclei are generated from several sources (Swift (1967)), including heated wires dipped in NaCl, AgCl or KI, carbon electrode arc, and spark production between tungsten electrodes at high voltage. The size and number of droplets produced by this generator depend on the concentration of nuclei, the boiler temperature, the rate of gas flow, and the dimensions of the condensation chimney.

The generation of aerosols containing droplets larger than 1 μm diameter requires low-nuclei and high-vapour concentrations. Swift (1967) used a Muir-type, Sinclair–LaMer system to show that aerosols with d_P values from 0.3 to 1.0 μm could be generated in a controlled manner by increasing the boiler temperature from 110 to 150 °C at constant nuclei concentration (8×10^5 nuclei cm^{-3}), with the particle mass being proportional to the vapour pressure of DEHS. Other tests were conducted at constant nuclei concentration (3×10^5 nuclei cm^{-3}), and revealed that the fraction of vapour condensing on the nuclei was greatest when the re-heater temperature was between 175 and 200 °C. Thus, d_P values increased sharply when the re-heater temperature was increased from 135 to 175 °C and remained constant in the range 175 to 200 °C, before decreasing at higher temperatures. Polydisperse aerosols were produced when the re-heater temperature was close to the boiler temperature (130 °C), but monodispersity improved at higher re-heater temperatures. Swift also found that d_P decreased as the gas flow rate was increased; droplets generated with a volumetric flow rate of 100 $cm^3 min^{-1}$ were three times larger than those obtained at 1000 $cm^3 min^{-1}$.

At very low nuclei concentrations ($< 10^4$ nuclei cm^{-3}), d_P is almost independent of the vapour concentration since nearly all condensation occurs on the chimney walls. Each nucleus is exposed to the same vapour concentration and grows to a fixed size (0.8 to 1.0 μm geometric diameter). At higher nuclei concentrations (10^5 to 10^7 nuclei cm^{-3}), there is a weak dependence of d_P on concentration, while at even higher concentrations ($> 10^7$ nuclei cm^{-3}) almost all of the vapour condenses on the nuclei to produce droplets with d_P values smaller than 0.4 μm. It is difficult to produce monodisperse aerosols with high nuclei concentrations because the coagulation rates of the nuclei and droplets become significant. Nevertheless, the upper limit of aerosol concentration available with this type of aerosol generator can be more than 100 times greater than that achievable with the VOAG or STAG.

The Rapaport–Weinstock aerosol generator consists of three stages. A DeVilbiss nebuliser generates a polydisperse mist of the high boiling-point liquid (DOP or DEHS) in the first stage, the coarse droplets are evaporated, and then they are re-condensed as a monodisperse aerosol. The condensation nuclei originate as non-volatile impurities in the liquid. This aerosol generator requires only a few minutes to attain thermal equilibrium in comparison with the Sinclair–LaMer generator which can take several hours to stabilise. Thermal decomposition of the liquid can be a problem in the Sinclair–LaMer generator, but is avoided in the Rapaport–Weinstock system because fresh liquid is continuously fed to the heated section. Furthermore, the temperature of the Rapaport–Weinstock vaporiser does not have to be kept constant; after evaporation of the droplets produced by the nebuliser, the vapour concentration ceases to be temperature dependent. Since the nuclei and vapour concentrations are proportional to the concentration of the nebulised droplets, changes in the latter do not appreciably affect the droplet size of the generated aerosol.

Aerosols with d_P values in the range 0.3 to 1.4 μm have been produced by changing the distance from the nozzle of the nebuliser to a special screen placed inside the nebuliser (Lassen (1960)). Limited changes in droplet size can also be achieved by introducing dilution air into the droplet stream entering the vaporiser.

The Monodisperse Aerosol Generator ((MAGE) Lavoro & Ambiente scrl, Bologna, Italy) is a condensation-type aerosol generator that has greater flexibility than the systems described above. This device produces monodisperse particles ($\sigma_g < 1.1$) in the size range 0.2 to 8 μm geometric diameter (Prodi (1972)). The MAGE has many similarities to the Sinclair–LaMer and Rapaport–Weinstock generators: a stream of nuclei is exposed to the vapour of a low-volatile liquid at an elevated temperature, and the controlled heterogeneous condensation of the vapour onto the nuclei results in the formation of the product aerosol. However, neither of the earlier systems are readily suited to the rapid, continuous production of solid particles as well as liquid droplets. The MAGE was designed to meet this need by improving the temperature control of the vessel containing the high-boiling liquid (bubbler vessel). Most significantly, the size of the particles can be adjusted to a new value in a few seconds by altering the proportion of the gas flow that bypasses the bubbler vessel.

The construction and operation of the MAGE has been described by Prodi (1972), and is shown in Figure 8. A Collison atomiser is used to generate nuclei in a nitrogen atmosphere from a dilute solution of a substance which is insoluble in the low-volatile liquid. Scheuch and Heyder (1986) have used aqueous solutions containing high-purity sodium chloride in the atomiser. Fluorescent tracer compounds (*e.g.* sodium fluorescein) have also been used to produce nuclei that can be measured when coated by the low-volatile liquid (Horton *et al* (1991)); these aerosols can be detected at very low mass concentrations by fluorimetry to minimise the time taken to calibrate aerosol analysers.

The droplets that leave the atomiser are dried as they pass through a vertical tube containing silica gel desiccant. This resulting nuclei stream is then split in two: part of the flow passes directly at ambient temperature to the reheater section (by-pass flow), while the remaining flow is bubbled through the vessel containing the low-volatile liquid (typically DEHS) by partly closing a needle valve located at the entrance to the bypass line. The bubbler vessel is maintained at sufficient temperature to produce a moderate flow of vapour without significant thermal degradation. Temperature control is typically maintained within $\sim 1\,^{\circ}$C to ensure a constant supply of vapour. On leaving the bubbler vessel, the vapour stream is immediately diluted with the flow from the bypass line before entering the reheater section to ensure the complete vaporisation of any remaining liquid. The reheater consists of a short, vertically-mounted resistance furnace containing a 10 mm bore glass tube through which the vapour–nuclei–gas mixture passes in the downward direction. Heterogeneous condensation occurs onto the nuclei at the outlet of the reheater where a well-defined thermal gradient exists from the reheater temperature to ambient temperature. The process is rapid when the vapour–nuclei–gas mixture is cooled to the point at which the vapour concentration exceeds the saturation

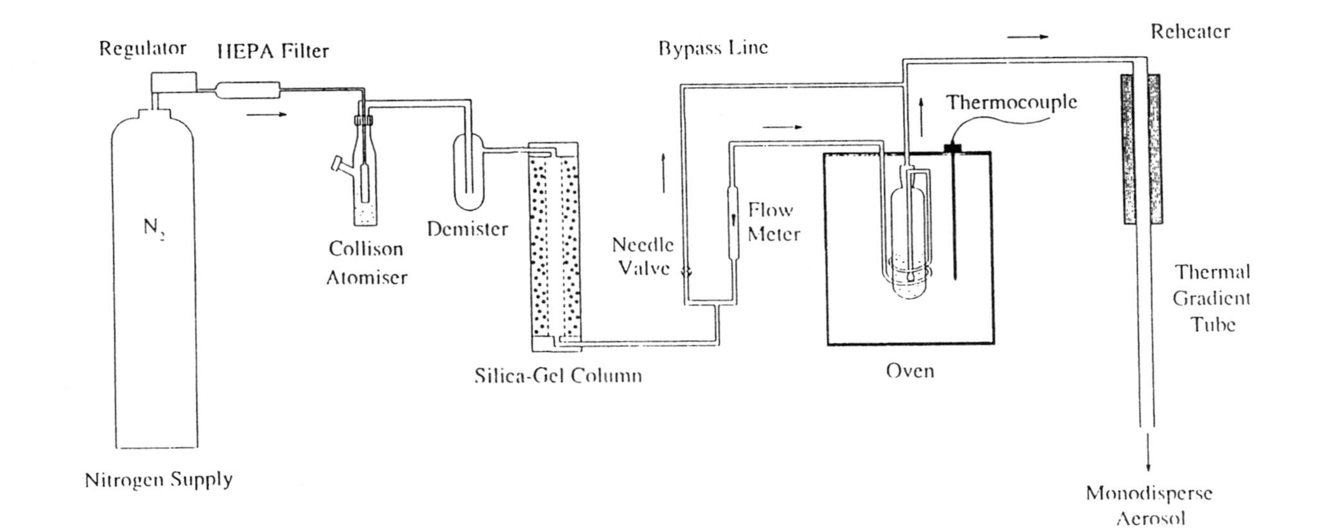

Figure 8 *MAGE monodisperse condensation aerosol generator*
(Reprinted by permission of Charles C. Thomas, Publisher, Ltd. from 'A Condensation Aerosol Generator for Solid Monodisperse Particles', by V. Prodi, p. 169 in 'Assessment of Airborne Particles', Eds: T. T. Mercer, P. E. Morrow and W. Stöber. Copyright 1972 by Charles C. Thomas, Publisher, Ltd.)

limit. The process results in uniform-sized droplets with an outlet concentration close to 10^5 droplets cm^{-3} in a flow rate of $3.5 l min^{-1}$.

The MAGE can be used to generate uniform crystals of materials, such as caffeine, that sublime in the bubbler vessel (Vaughan (1990)). Vapour condensation takes place directly onto the nuclei from the vapour to the solid phase, and the resulting particles have the crystal habit of the low-volatile substance. These particles are a useful supply of particle shape standards.

Horton *et al* (1991) have undertaken a laboratory-based evaluation of the MAGE:

(a) Incomplete drying of the nuclei (NaCl) does not have any significant impact on aerosol monodispersity (DEHS) if the reheater temperature is sufficiently high to ensure complete evaporation.
(b) Premature condensation of the low-volatile vapour between the bubbler vessel and reheater must be avoided by trace-heating this section of the MAGE.
(c) The reheater temperature should be at least 50 °C higher than the temperature in the bubbler vessel, and the establishment of optimum operating conditions for a new material is best achieved by experimental studies.
(d) The diameter of the aerosol droplets produced by the MAGE is affected by the amount of liquid in the bubbler vessel. Reproducible aerosols were produced when the depth of liquid was maintained within 0.5 cm of 5.0 cm.
(e) Adjustments to the ratio of bubbler vessel to bypass flow rates resulted in an almost immediate change in the size of the aerosol produced by the MAGE. Stable conditions (at the new size) were achieved within one minute of any alteration in the flow conditions.

Other condensation aerosol generators have been described by Nicolaon *et al* (1970) and Fuchs and Sutugin (1963).

Electrostatic Classifier

The electrical mobility of a particle (Z_p in $cm^2 V^{-1} s^{-1}$) is a unique function of the diameter (d_p in μm) given by:

$$Z_p = \left[\frac{3 \times 10^{15} Ce}{\pi \mu d_p}\right]$$

for particles carrying one elementary unit of electric charge, where e is the electronic charge constant (1.60×10^{-19} C), C is the Cunningham slip correction factor, and μ is the gas viscosity (dyn s cm^{-2}). An electrostatic classifier described by Liu and Pui (1974) is used as a primary standard for producing calibration particles of known size, since the electrical mobility–particle size relationship can be determined precisely by this equation. An example of a commercially available electrostatic classifier is that manufactured by Thermo-systems Inc. (TSI), St Paul, Minnesota, which is based on electrical mobility and can produce a monodisperse aerosol by precipitating all particles selectively except those within a narrow size range. A pneumatic nebuliser is used to generate a polydisperse aerosol which is dried and brought to a state of charge equilibrium with bipolar ions by means of a krypton-85 radioactive source. The charges on the particles obey a Boltzmann distribution, and most particles are electrically neutral or carry a single positive or negative charge. This aerosol is introduced in laminar flow to the classifier (sometimes referred to as a differential mobility analyser (DMA)), which is a cylindrical vessel containing concentric cylindrical electrodes. The inner electrode is maintained at a high voltage and the outer electrode is maintained at earth potential. Charged particles in the aerosol stream flow along the outer tube and are deflected by virtue of their electrical mobility. If a negative potential is applied to the central electrode, positively charged particles move across the core of clean air before arriving at the rod. Other particles are either unaffected because they are neutral or are deflected to the outer electrode. In either case, they are swept to a filter in the air flow through the classifier. Only particles within a narrow band of electrical mobility have trajectories which enable them to arrive at the exit slot at the base of the centre electrode, where they emerge from the classifier as a reasonably monodisperse aerosol.

The TSI model 3071 classifier has a useful range from 0.01 to 1.0 μm diameter, producing monodisperse aerosols that have σ_g values in the range 1.04 to 1.08 under the best operating conditions. However, the size distribution of the original polydisperse aerosol must be carefully selected so that the modal size is smaller than the desired size of monodisperse particles. If this precaution is not observed, the dispersity of the aerosol leaving the classifier will be increased by additional particles twice the desired size carrying double the electrical charge (*i.e.*, same electrical mobility). The problem of multiple-charging becomes more severe as the particle size increases (Fuchs (1963)).

As well as providing a source of monodisperse particles, the classifier can be used to calibrate Condensation Particle Counters (see Section 5.8), since almost all the particles carry the same charge that can be measured in an electrometer. The electric current (I in A) is given by:

$$I = QeN$$

where N is the particle number concentration (particles cm^{-3}), and Q is the volumetric flow rate into the electrometer (cm^3 s^{-1}).

Other Sources of Monodisperse Particles

Several species of pollen are commercially available (Polysciences Inc, Warrington, PA, USA) that are highly monodisperse, with d_P values in the range 5 to 30 μm and σ_g values between 1.04 and 1.07 (Dennis (1976)). A limited number of pollens such as clover are spherical, with rough surfaces. A species of orchard grass (*Dactlyis glomerata*) has elliptical pollen grains, while certain types of fungi produce monodisperse spores: d_P values for Lycoperdon and Lycopodium spores are 2.09 and 15 μm respectively, with σ_g values in the range 1.08 to 1.10. Dry dispersion methods as described later in this document are used to make these particles airborne.

Bacteria, red blood cells and inorganic powders prepared under carefully controlled conditions have become available as sources of monodisperse aerosol. Titanium dioxide particles (mostly in the anatase form) with a mean d_P of 0.6 μm and σ_g of 1.09 are available in aqueous suspension from Polysciences Inc. This material is formed by high-temperature hydrolysis of aerosol droplets of titanium tetrachloride (Matijevic *et al* (1977)). Similar sized TiO_2 particles have also been made by the controlled hydrolysis of dilute alcoholic solutions of titanium alkoxides (Barringer and Bowen (1982)). Haggerty and Cannon (1981) have prepared submicron monodisperse ceramic powders (alumina) by laser-driven gas-phase reactions at extremely high temperatures.

Particle shape standards refer to non-spherical particles that have well-defined shapes. There are no formally certified shape standards at present, although non-spherical particles with regular 3-D shapes have been used to assess the performance of various types of aerosol analyser. While the majority of techniques used to determine the size of aerosol particles relies on operating principles that assume that the particles are spherical, demands for shape standards are growing as the aerosol measurement community realises the importance of calibrating measurement techniques with reference particles appropriate to their intended application, which frequently involves the monitoring of relatively complex shapes (Lewis *et al* (1993) and Marshall (1995b)). Therefore, a series of three fibre-analogue shape standards have been prepared as part of the VAM project (see Appendix B).

4.1.2 Polydisperse Aerosols

Compressed Air Nebulisers

The simplest way to generate a droplet aerosol is by means of a pneumatic nebuliser, in which an aerosol of small droplet size is produced by removing the larger droplets via impaction before they leave the device. The resulting aerosol is relatively polydisperse (σ_g values are typically between 1.5 and 2.0), with a sharply defined cut-off at the upper end of the size distribution. Nebulisers produce aerosols with mass concentrations from 5 to 50 mg m^{-3}, and mass median diameters (MMD) in the range 1 to 10 μm. The operating principle of most pneumatic nebulisers is very much the same: compressed air at a pressure of 35 to 270 kPa (5 to 40 psig) enters the DeVilbiss type-40 nebuliser at high velocity from a small-bore tube. The low pressure created by the Bernouilli effect causes liquid to be drawn from the reservoir through a second tube into the airstream. This liquid emerges as a thin filament which ruptures into droplets when accelerated in the airstream. The coarse spray is impacted on the wall of the aerosol outlet tube, where the large droplets are deposited and drain back to the reservoir. Compressed air escapes from the Babington nebuliser (Babington *et al* (1969)) at high velocity from a small slit or hole in a hollow sphere over which liquid flows to form a film that is shattered by the flow of compressed air to generate the aerosol.

Most nebulisers produce a maximum number concentration in the range 10^6 to 10^7 droplets cm^{-3}. Pneumatic nebulisers may be used with pure liquids or with dilute solutions in a volatile solvent (*e.g.* NaCl in water). In the latter system, the resulting aerosol contains mainly sub-micron particles and may be used as the starting material for producing a monodisperse aerosol by means of the electrostatic classifier as described earlier.

Ultrasonic Nebulisers

The dispersing force in an ultrasonic nebuliser is the mechanical energy produced by a piezoelectric crystal vibrating in an electric field induced by a high frequency oscillator. These generators can produce a higher concentration of aerosol than achieved with pneumatic nebulisers, and the mass median diameter of the particles is generally small ($<1\,\mu m$). The sizes of droplets formed in the ultrasonic nebuliser designed by Raabe (1976) depend on the frequency of the acoustic field and the degree of coupling between the piezo-electric crystal and the liquid being nebulised. Droplet coagulation may be a problem because of the high initial particle concentrations.

Fluidised Bed Aerosol Generators

Although not strictly used to calibrate aerosol analysers, polydisperse aerosols generated under controlled conditions from a fluidised bed generator can be useful in monitoring the response of aerosol analysers. The fluidised bed aerosol generator is one of the most convenient of several methods for the dry dispersion of dusts, with output concentrations in excess of $100\,mg\,m^{-3}$. The technique has been reviewed by Guichard (1976), and an example of such a generator is that based on the design of Marple *et al* (1978). A similar aerosol generator is commercially available (TSI model 3400 from Thermosystems Inc., St Paul, Minnesota), with the layout shown in Figure 9. The aerosol generator designed by Marple consists of a 5.1 cm diameter fluidised bed filled with $180\,\mu m$ diameter bronze beads to a depth of 1.5 cm. Dust is eluted directly from the fluidised bed into a chamber containing a krypton-85 radioactive source to achieve Boltzmann charge equilibrium. The commercial systems use $100\,\mu m$ diameter beads, and contain a cyclone at the outlet of the elutriator to remove any agglomerated particles. Particles as small as $0.5\,\mu m$ and as large as $50\,\mu m$ diameter can be suspended by this technique. The test dust is metered into the fluidised bed by a conveyor feed, and a constant output is achieved after several hours operation to give a well-defined source of aerosol.

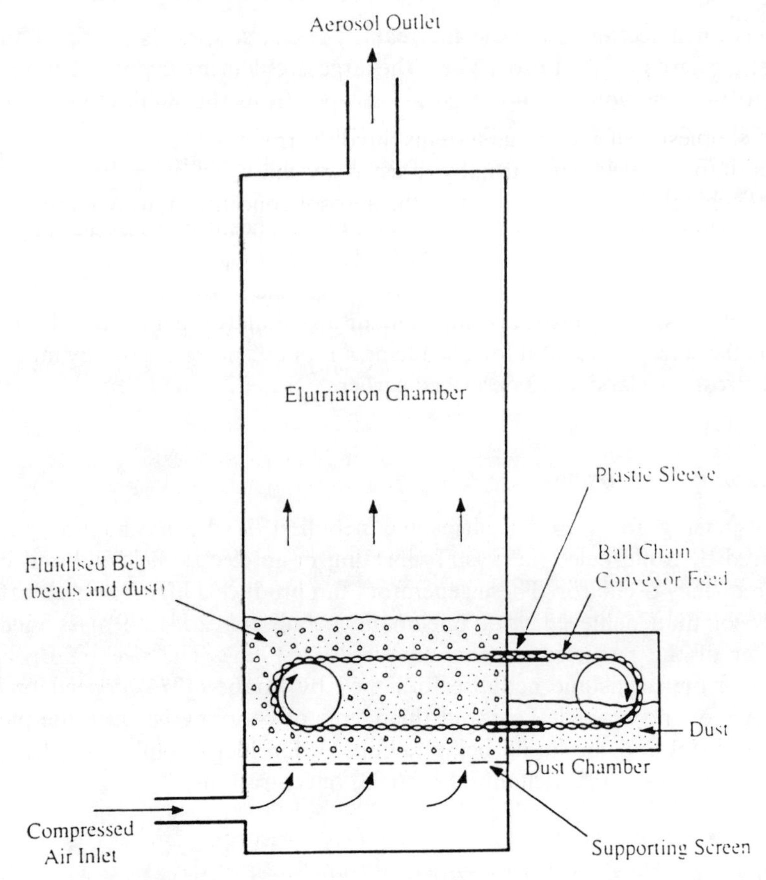

Figure 9 *Fluidised-bed aerosol generator*
(Reprinted by permission of American Industrial Hygiene Association from 'A Dust Generator for Laboratory Use', by V. A. Marple, B. Y. H. Liu and K. L. Rubow, *Am. Ind. Hyg. Assoc. J.*, **39**, 26. Copyright 1978 by American Industrial Hygiene Association.)

Other Dust Generators

A wide variety of methods exist for the dispersion of dry powders to give aerosol mass concentrations varying from less than one to more than $100\,\mathrm{mg\,m^{-3}}$. Many of these systems are described by Dennis (1976), and therefore only the most important aspects of their use are presented here. A number of useful references describe the operation of this type of generator, including Wright (1950), Craig *et al* (1972), Drew and Laskin (1971), Fuchs and Murashkevich (1970), Hounam (1971), Knutson *et al* (1967) and Cadle and Magill (1951). Dust generators are not often used in calibration studies because they are not as well-defined and controllable as other techniques.

The basic requirements for a dust generator are:

(a) continuous metering of the powder feed,
(b) dispersion of the powder to form the aerosol.

The simplest self-metering systems involve the gravity feeding of loose powder into an airstream, usually assisted by vibration, to give an uneven delivery which causes fluctuations in the aerosol concentration. A more stable dust metering system is provided by a cylinder of compressed powder which is eroded or scraped away at a constant rate (Wright (1950)).

Powder dispersibility depends on both the physical and chemical properties of the material, including the particle size distribution, shape and moisture content, and the generation of electrostatic charge. Incomplete dispersion results in an aerosol particle size distribution which is larger than that of the original powder; hydrophobic materials such as talc are easier to disperse than hydrophilic substances. The most common method of dust dispersion is to feed the powder into a high-velocity airstream so that the shear force in the turbulent flow results in de-agglomeration and efficient dispersion.

Electrostatic charge is a problem common to all dry-dispersion dust generators, which is exacerbated when dry powders are suspended in conditions of low humidity. Excessive charging results in extensive deposition on nearby surfaces, which reduces the aerosol concentration. It is advisable to pass the aerosol as soon as possible after dispersion through an electrostatic charge equilibrator.

A size-selective classifier can be introduced at the outlet of the dust generator to remove unwanted agglomerates and particles larger than the chosen size. Cyclones, impactors, elutriators and sedimentation chambers have also been used to modify the size distribution of powder aerosols. This is particularly important in the calibration of respirable dust monitoring devices, when the test aerosol should meet the size requirements defined in terms of the health-related aerosol fractions as a function of aerodynamic diameter (Vincent (1989) and ISO (1981)).

Both the TSI Small-Scale Powder Disperser (SSPD (model 3433)) and the Palas Dry Powder Disperser (DPD (model RBG-100)) are commonly used to generate dry powder aerosols (see Figures 10 and 11, respectively). The SSPD makes use of the suction force generated when an air stream is expanded through a Venturi to lift particles from a collection substrate such as a filter (Blackford and Rubow (1986)). Suspended aerosol is deagglomerated by the strong shear force encountered as the particles pass through the Venturi section prior to sampling. The SSPD is most useful with dry powders and is only capable of resuspending small amounts of material (3 to 90 mg h^{-1}), resulting in aerosol mass concentrations between 3 and 100 mg m^{-3}. It is highly effective at resuspending large-diameter polymer latex particles ($> 5\ \mu m$ diameter) that are not readily produced as an aerosol by atomisation/nebulisation. Particles in the size range 1 to 50 μm diameter can be easily suspended by the SSPD; smaller particles tend to remain agglomerated after suspension and larger particles are too heavy to suspend by the capillary aspiration technique.

The DPD operates by forcing a plug of powder into the path of a rotating

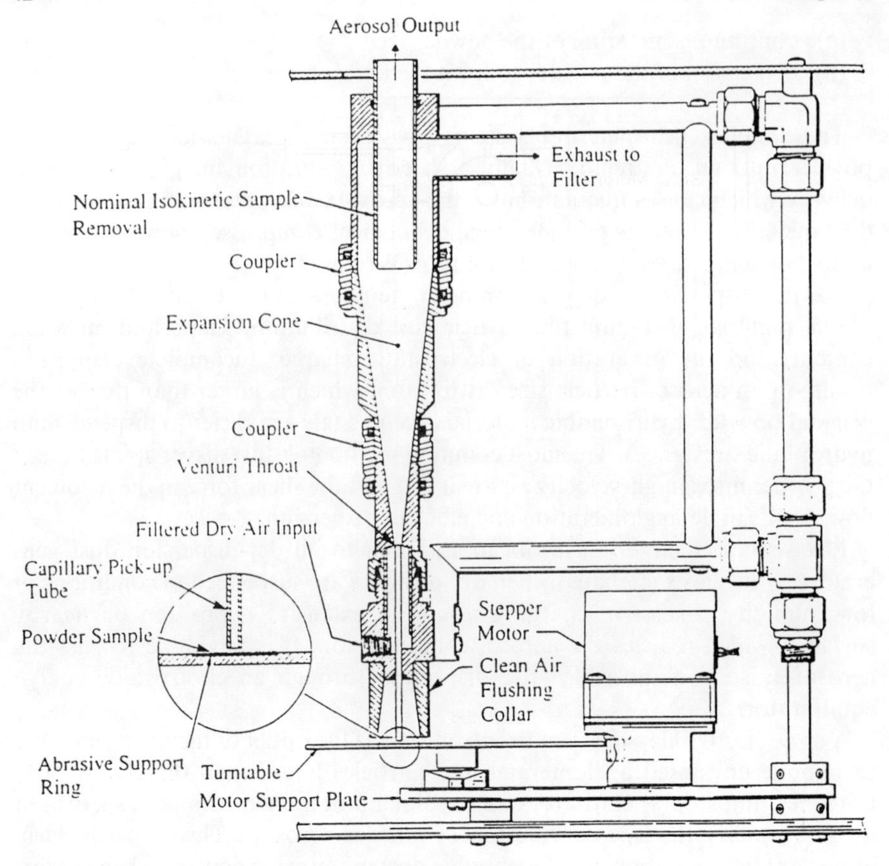

Figure 10 *TSI small-scale powder dispenser (SSPD)*

wire-brush (Zahradnicek and Löffler (1976)). Loosened powder is rapidly dispersed from the rear of the generator as an aerosol in a flow of clean air (Figure 11). The rate of removal of powder from the compacted mass is proportional to the product of the feed rate and the cross-sectional area of the compacted powder (piston diameter), and is independent of the properties of the powder as long as the material flows freely. Feed rates can be varied between 40 and 400 g h^{-1} to produce aerosol mass concentrations in the range 0.005 and 50 g m^{-3}. The technique is very effective for wind-tunnel calibrations of aerosol analysis equipment when the aerosol is highly diluted by the large volumetric flow.

4.2 Flow Calibration

Calibrations are generally carried out to establish the relationship between instrument response and reference values of the parameter being measured. Aerosols are defined as metastable suspensions of liquid droplets and solid particulate in a suitable flow of gas, and therefore the flow rates of gasborne

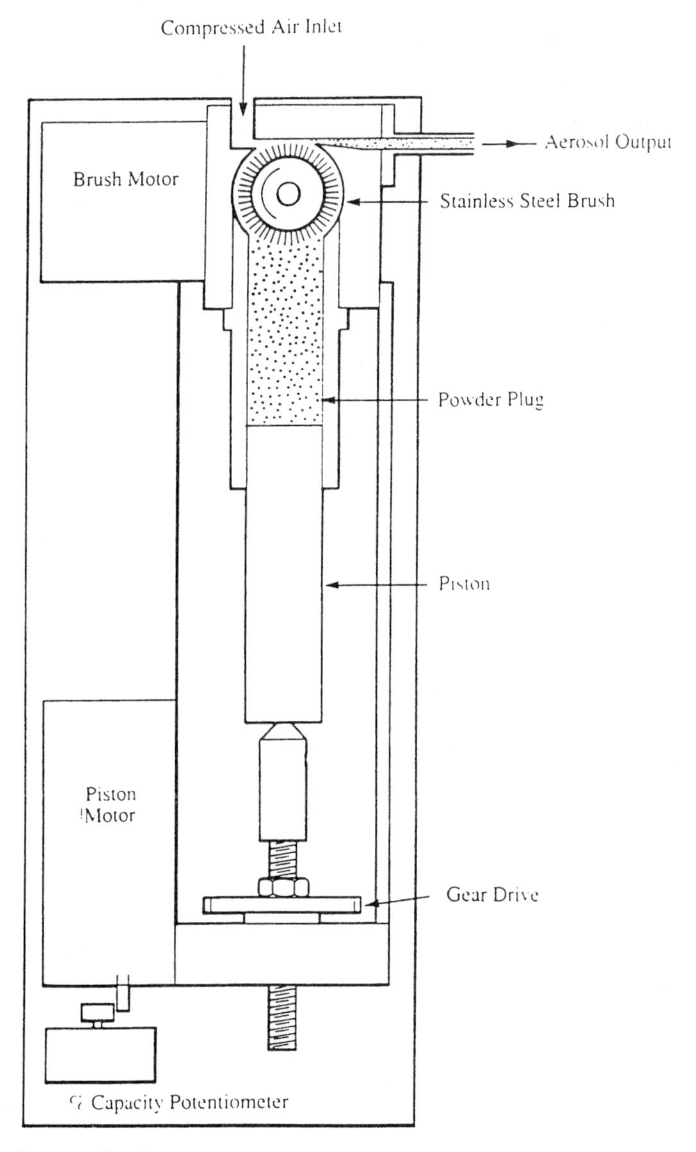

Figure 11 *Dry powder dispenser*

aerosols constitute an important parameter in the calibration of the sampling characteristics of an aerosol analyser. The NIOSH Manual of Analytical Methods recommends that sampling pumps should be regularly calibrated with use, and that any such calibration exercise be carried out with the sampler in-line (NIOSH (1984)). All calibration data should be recorded with each unit. These general recommendations are equally applicable to aerosol analysers, and therefore flow calibrations need to be checked according to the manufacturer's instructions.

4.2.1 Flow Rate Metering Instruments

Accurate measurements of air flow and volume form an integral part of the calibration of the vast majority of aerosol analysers, although size analysis methods based on microscopy and electrical sensing-zone techniques are obviously exceptions to this requirement.

Flow rate metering is divided into two general categories: primary and secondary standards. Primary measurements involve the direct measurement of volume on the basis of the physical dimensions of an enclosed space. Secondary standards are reference instruments or meters calibrated in a traceable manner to primary standards, and shown to be capable of maintaining their accuracy with reasonable handling and care. Different types of flow meter, the parameter measured, and the range of flow rate are given in Table 4.2. Both primary and secondary flow standards are presented below.

(a) Primary Standards

Spirometer. A spirometer is a cylindrical bell supported by a chain and balanced by a counter weight, with the open end under a liquid seal. The pressure exerted by the bell, and therefore the resistance to movement as air moves in and out of the bell, is kept constant by the cycloid counterpoise which compensates automatically for buoyancy changes exerted by the liquid. The volume of air entering the spirometer is determined by calculating the change in height of the bell and multiplying by the cross-sectional area.

The Mariotte bottle (a type of aspirator) is a similar instrument to the spirometer which measures displaced liquid instead of air. The volume of air

Table 4.2 *Equipment for flow rate calibration of aerosol size analysers*

Meter	*Quantity measured*	*Flow rate*
Spirometer (P)	Integrated volume	6 to 600 l min^{-1}
Soap film flow meter (P)	Integrated volume	2 to 10 000 ml min^{-1}
Mercury sealed piston (P)	Integrated volume	1 to 12 000 ml min^{-1}
Wet test meter (S)	Integrated volume	Unlimited volumes, max flow rates 0.5 to 230 l min^{-1}
Dry test meter (S)	Integrated volume	Unlimited volumes, max flow rates 10 to 150 l min^{-1}
Electronic mass flow meter (S)	Mass flow rate	0 to 10 ml min^{-1} up to 0 to 15 000 l min^{-1}
Laminar flow meter (S)	Volumetric flow rate	0.02 ml min^{-1} to 1 m^3 min^{-1}
Venturi meter (S)	Volumetric flow rate	Depends on pipe and orifice diameters
Orifice meter (S)	Volumetric flow rate	Depends on pipe and orifice diameters
Rotameter (S)	Volumetric flow rate	From 1.0 ml min^{-1} upwards
Thermo-anemometer (S)	Velocity	From 0.3 m min^{-1} upwards
Pitot tube (P)	Velocity	From 300 m min^{-1} upwards

(P) = Primary Calibration Standard; (S) = Secondary Calibration Standard

drawn into the bottle is equal to the change in liquid level multiplied by the cross-sectional area at the top of the water surface.

Frictionless Piston Meter. Cylindrical air displacement meters with nearly frictionless pistons are often used for primary calibrations at flow rates in the range 1 to 10^3 ml min^{-1}. The simplest of these systems is the soap bubble meter in which the movement of a film bubble is timed as it is constrained by the air flow to traverse a known volume of a burette.

Frictionless piston meters can measure flow rates to an accuracy of 1%, but this value decreases as the flow rate increases and air permeates the soap film. The basic instrument is easy to construct, and the passage of the soap film can be timed manually with a stop-watch. There are also more sophisticated devices with automatic read-out that are available commercially, *e.g.* the Gilibrator (range with multiple elements between 0.2 and 30 l min^{-1}) and DryCal instruments (BIOS International).

Pitot-Static Tube. The spirometer and frictionless piston are primary standards for the measurement of volume and flow rate, while the Pitot-static tube is the primary standard for measuring gas velocities. A Pitot-static tube consists of an impact tube which faces axially into the flow, and a static area formed by a concentric tube with eight holes placed equally and axially around the tube in a plane which is eight diameters from the impact opening (Figure 12). The difference between the static and impact (total) pressure is the velocity pressure, and Bernoulli's theorem is used to calculate the gas flow.

The accuracy of a Pitot tube is limited by the ability to measure the velocity pressure. Above 12.7 m s^{-1}, a U-tube manometer can be used, but for lower velocities an inclined manometer or a low-range Magnehelic pressure gauge is required.

(b) Secondary Standards

Secondary standards are reference instruments which trace their calibration to primary standards. Amongst such standards are a number of instruments which provide an accuracy comparable to that of the primary standards, but which cannot be calibrated by internal volume measurement. These instruments are therefore sometimes called intermediate standards and provide an accuracy of $\sim$1%.

Wet Test Meter. A wet test meter consists of a cylindrical container in which a partitioned drum is half submerged in water, with openings at the centre and periphery of each radial chamber. Gas enters at the centre and flows into an individual compartment causing it to rise, and thereby induce rotation which is dependent on the fluid level in the meter, since the liquid is displaced by air. This liquid level must be maintained at a calibrated height indicated by a sight-gauge. Once the instrument is filled with water, the meter should be saturated with the gas in question by passing the gas through the instrument for several hours.

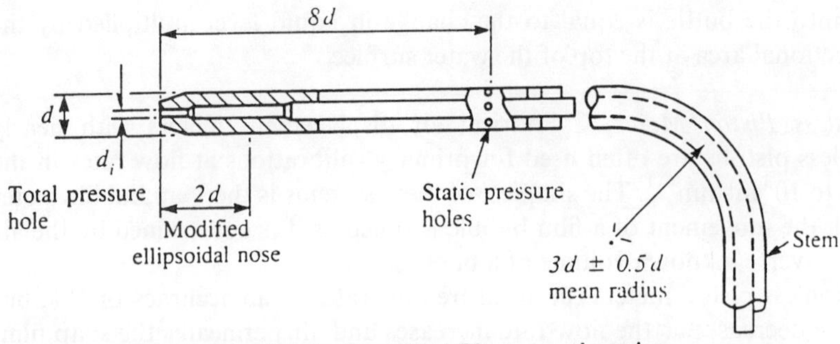

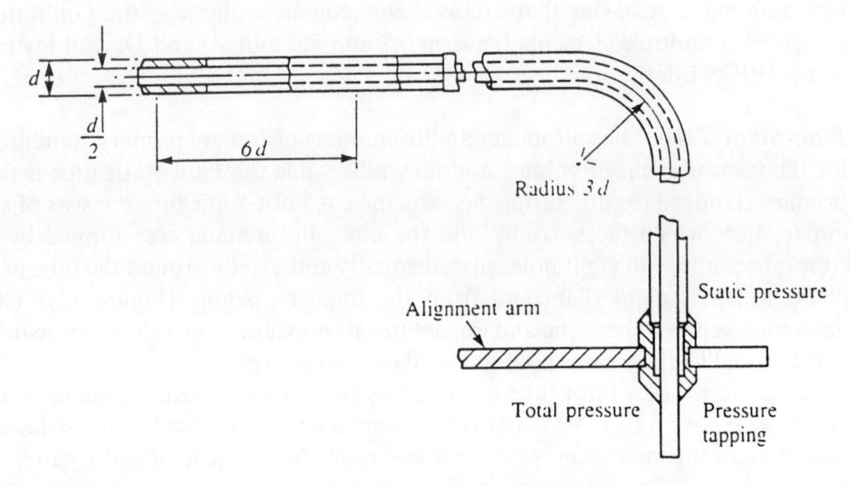

Figure 12 *Pitot-tube designed in accordance with BS 1042 : 1984*

When calibrated against a spirometer, wet gas meters exhibit an accuracy of 0.5% or better.

Dry Gas Meter. A dry gas meter is very similar to the instrument used to meter domestic gas supplies. Two bags are interconnected by mechanical valves and are linked to a cyclic counting device. Air or gas fills one bag while the other empties; when the cycle is completed, the valves are switched, and the second bag fills while the first one empties. At pressures up to 1717 kPa, an accuracy of ~1% can be achieved, and this value can be improved by calibrating against a spirometer.

(c) Additional Secondary Standards

While the positive displacement meters described above consist of a tight-fitting moving element with individual volume compartments that fill at the inlet and discharge at the outlet ports, other secondary standards are based upon the conservation of energy and utilise Bernoulli's theorem for the exchange of potential energy and/or friction heat. A flow restriction within a closed conduit causes an increase in kinetic energy, which requires a corresponding decrease in potential energy (*i.e.* static pressure). The flow rate can be calculated from a knowledge of the pressure drop, the flow cross-sectional area at the restriction, the density of the fluid, and the ratio of the actual to theoretical flow. Flow meters which operate on this principle can be divided into two groups:

(a) variable-area meters (including rotameters) in which a constant differential is maintained by varying the flow cross-sectional area, and
(b) variable-head meters (including Venturi meters and flow nozzles) that have a fixed restriction in which the differential head varies with the flow.

Rotameters. A rotameter consists of a 'float' that is free to move up and down within a vertical tube which is larger at the top than the bottom. The fluid flows upwards, causing the float to rise until the pressure differential across the annular area between the float and the tube wall is just sufficient to support the float. Rotameters come in a range of different sizes, and can be used in a flow range from $\sim 2 \times 10^{-6}$ ($2\,\mathrm{ml\,min^{-1}}$) to $3\,\mathrm{m^3\,min^{-1}}$. One or more rotameters often form an integral part of the flow measurement circuitry for many types of aerosol analyser.

Rotameters supplied with a calibration curve by the manufacturer have nominal accuracies of $\pm 5\%$ that can be improved to between $\pm 1\%$ and $\pm 2\%$ when calibrated in the operational measurement system. Most rotameters are calibrated against a primary standard or a more accurate secondary standard. These calibrations are usually carried out with one of the rotameter ports open to the atmosphere. Accuracy is therefore limited by the reproducibility or correction of these conditions: if one side of a rotameter is not open to the atmospheric pressure, the calibration as supplied is no longer valid. When used in a measurement system such as an aerosol sizing instrument, rotameters are very often set up so that neither port is open to atmospheric pressure, and therefore must be calibrated *in situ* against a known standard. This arrangement occurs frequently when factory-calibrated rotameters are fixed 'in-line' in many 'home-made' aerosol samplers, and is a common cause of error in flow measurements.

Orifice Meter. The simplest and least expensive variable-head meter is the square- or sharp-edged orifice. This type of meter can be calibrated against a reliable reference instrument so that the relationship between the measured pressure drop across the orifice can be established against the air flow rate. Manometers are used to monitor the pressure drop, and hence the flow rate.

Venturi Meter. The large energy loss of an orifice is caused by the sudden increase in area after the air has passed through the orifice restriction. Venturi meters have optimum converging and diverging angles from the orifice of 21°

and 5° to 15°, respectively. Their range of operation is extensive, and depends on the size of the pipe and orifice diameter.

Laminar Flow Meter. The drop in pressure is directly proportional to the flow rate in a laminar flow meter. Commercial instruments have restrictors that consist of egg-box or tube-bundle arrays of parallel channels, although simpler systems can be made with a 'T'-connection projecting down into a cylinder filled with oil or water to measure flow rates as low as 0.02 ml min^{-1} (2×10^{-8} m^3 min^{-1}).

Thermo-anemometer. A heated-element anemometer operates on the principle that a flow of air cools the sensor in proportion to the velocity of air over the range 0.05 to 40 m s^{-1}. These instruments are non-directional, measuring the airspeed but not the direction, and a wide range of such devices is available.

4.2.2 Calibrating Flow and Volume Meters

Various methods are used to calibrate the wide range of flow and volume meters. Comparisons should be made at well-defined time intervals (say every six months) between the performance of primary and secondary standards to maintain a growing confidence in all subsequent measurements with the secondary system. The reader is referred to a comprehensive text in which the calibration procedures are described for these instruments (Lippmann (1989)), along with the references cited therein; while essential for aerosol concentration measurements in particular, airflow calibrations do not constitute the main thrust of this document towards aerosol particle size analysers.

CHAPTER 5

Aerosol Particle Size Analysers—Calibration Procedures

5.1 Inertial Techniques

Inertial analysers can be conveniently split into three different types:

(a) impactors, impingers and cyclones,
(b) various spectrometers which operate by inertial separation, and
(c) 'real-time' instruments (*e.g.* TSI Aerodynamic Particle Sizer).

These inertial systems measure the aerodynamic rather than the physical characteristics of an aerosol ensemble, and therefore it is important to use calibrant aerosols that have well-defined aerodynamic properties.

5.1.1 Impactors, Impingers and Cyclones

All three devices fractionate the aerosol into discrete size ranges, and the object of a calibration is to determine the aerodynamic diameters at which each stage collects particles with an efficiency of 50% (d_{50}). The calibration methodology comprises the following steps:

(a) selection of calibrant particles,
(b) instrument preparation,
(c) generation, conditioning and characterisation of calibrant aerosol,
(d) sampling of calibrant aerosol,
(e) quantitative measure of material collected,
(f) data analysis and reporting.

Guidance on each of these steps is given below. It is recommended that calibrations for inorganic-based aerosols are performed on a mass basis, as this approach is the mode of operation in which the devices will ultimately be used.

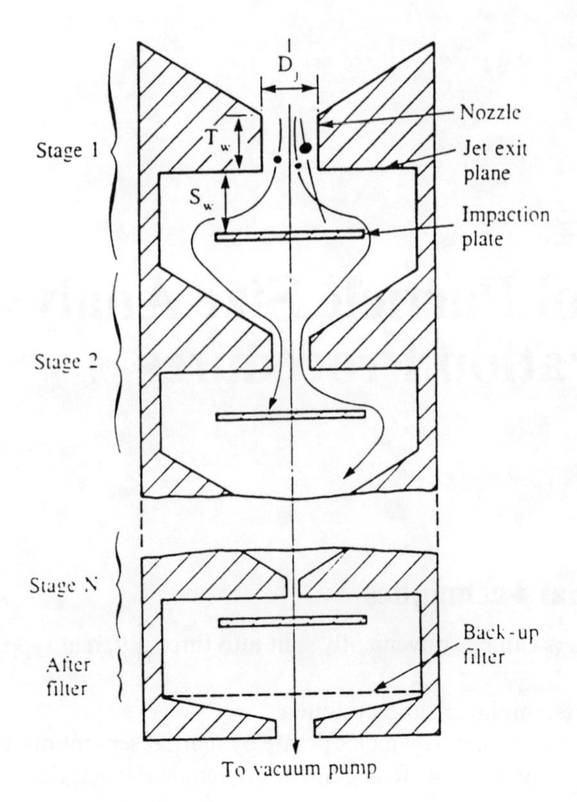

Figure 13 *Principle of cascade impactor*

When used to study bioaerosols, quantification is defined in terms of the number of micro-organisms (rather than mass); but additional uncertainties exist in the ability to detect and count all particles with 100% efficiency. Although number-based calibrations can be carried out with the equipment incompletely assembled (*i.e.* without all stages present), we recommend that the calibration is performed on a fully assembled device.

Several impactor stages are operated in series for size distribution analyses (Figure 13 illustrates the operational principle of a cascade impactor). Impactors can be grouped into four main types (ambient, in-stack, low-pressure and virtual systems), and examples are shown in Figure 14. Each stage must be calibrated, and this procedure can be time-consuming for systems that contain as many as six or more stages. Although many users rely on the calibration curves provided by the manufacturer, this practice should be avoided because even minor changes to the geometry can have a marked effect on the performance characteristics, especially for multi-orifice impactors. It should also be noted that almost all of the operational requirements for impactor calibration apply equally to cascade cyclone trains (as shown in Figure 15); therefore, cyclones are only mentioned specifically when it is necessary to point out some distinguishing feature of their performance compared with impactors.

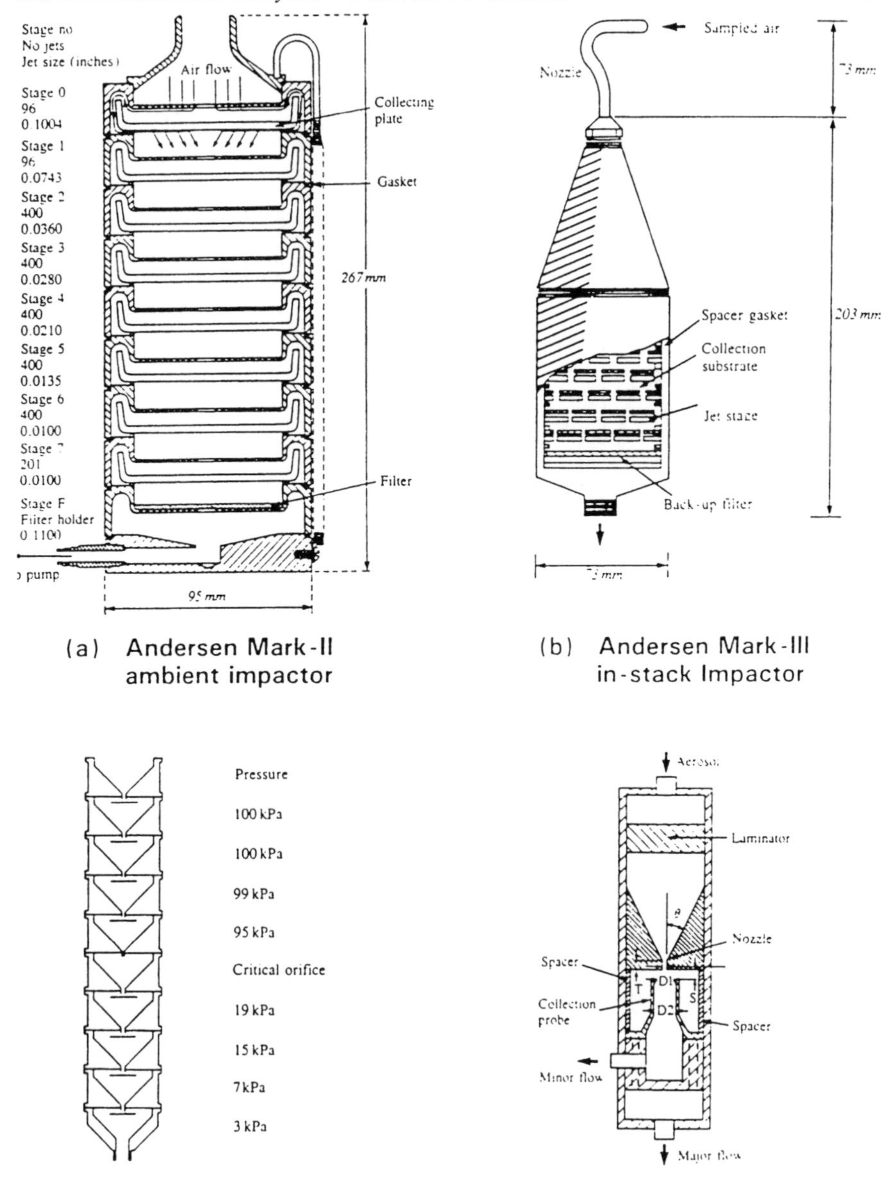

(a) Andersen Mark-II ambient impactor

(b) Andersen Mark-III in-stack Impactor

(c) Hering "low-pressure" impactor

(d) Chen "virtual" impactor (single-stage shown)

Figure 14 *Various types of impactor (continued on p. 52)*

(a) Selection of Calibrant Particles

Calibrations can be performed using either monodisperse or polydisperse particles. The advantage of monodisperse particles is that the test particles are well defined in terms of size, shape, density and optical properties. Although

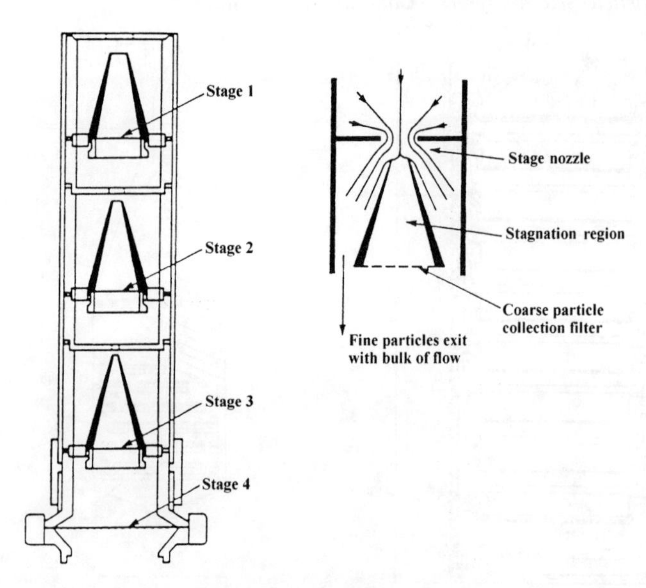

(e) UKAEA virtual impactor

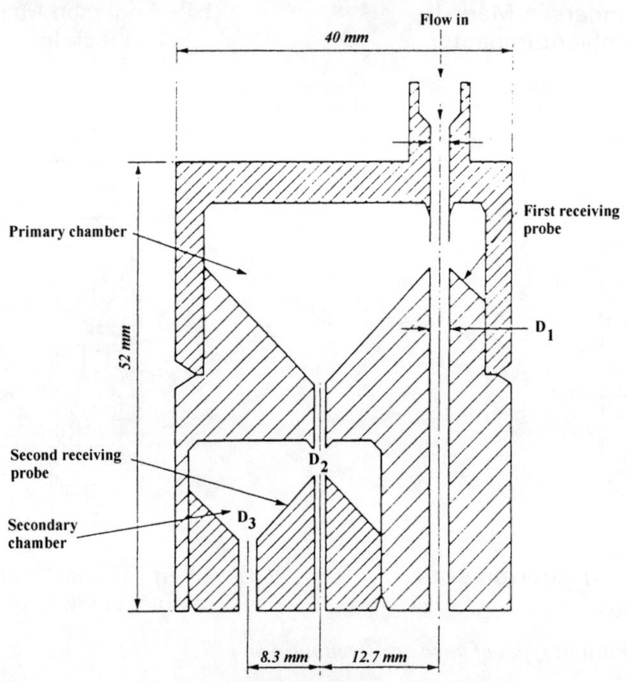

(f) Novick and Alvarez miniature virtual impactor

Figure 14 *Continued*

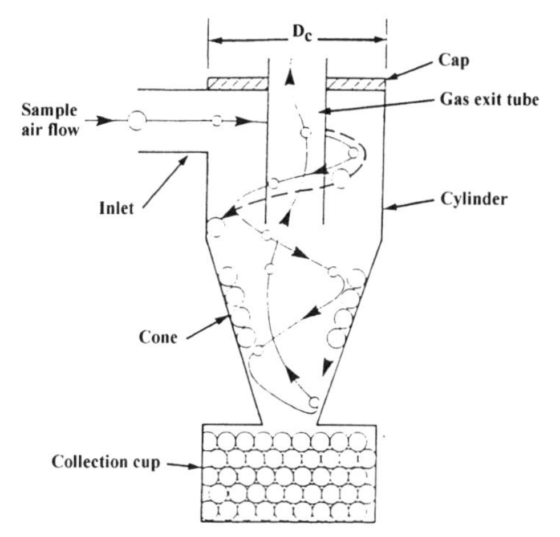

(a) Hypothetical flow through typical reverse flow cyclone

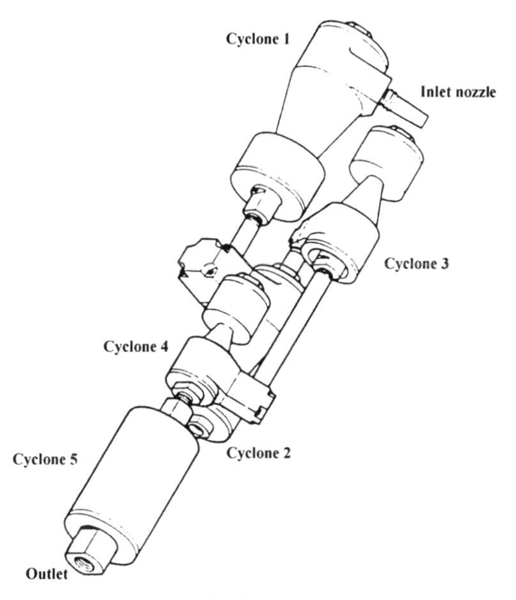

(b) Sierra-Andersen cyclone train

Figure 15 *Reverse-flow gas cyclones: operating principle and example of a cascade cyclone train*

several experiments must be performed to cover the operating range of the device, these tests can be minimised by careful selection of the particles' sizes. Polydisperse particles have the advantage that the complete system can be calibrated in one experiment, although data analysis is more complicated than with monodisperse particles and some uncertainties exist in demonstrating traceability to the international metre. Calibrations using monodisperse particles are recommended for the greatest accuracy. Thought must also be given to the composition of the calibration particles so that the mass collected on each stage can be easily determined using a traceable technique. It is recommend that either polymer latex particles are used with a fluorescent dye dispersed uniformly throughout the polymer matrix for particles smaller than 3.0 μm aerodynamic diameter, or monodisperse particles of a chemical dye (*e.g.* methylene blue) are generated by a vibrating orifice aerosol generator for larger particles (VOAG, see Section 4.1.1).

Detailed descriptions of calibrations using monodisperse aerosols have been given by Flesch *et al* (1967), Lundgren (1967), O'Connor (1973), May (1975), Smith *et al* (1979), Franzen *et al* (1978 and 1979) and Wang and Libkind (1982). Calibrations with polydisperse particles have been reported by May (1945), Mercer and Chow (1968), and Fairchild and Wheat (1984).

(b) Instrument Preparation

It is essential that the operating instructions provided by the manufacturer of the device under calibration are closely followed with respect to instrument preparation. Attention should be given to ensuring that the device is clean and correctly assembled. Collection media should be appropriately prepared, for example, filters inserted correctly, impactor plates greased if appropriate, or impinger stages loaded with the correct volume of liquid. The device must be leak tight so that the flow rate measured at the inlet corresponds to the flow rate throughout the remainder of the instrument.

(c) Generation, Conditioning and Characterisation of Calibration Aerosol

Preformed polymer latex particles are easily aerosolised using conventional pneumatic medical nebulisers (see Section 4.1.1). Precautions should be taken to ensure that the concentration of the latex suspension within the nebuliser reservoir is appropriate to minimise the number of multiplet particles produced (doublet, triplet, etc). Chemical dye particles are best generated using a vibrating orifice aerosol generator (see Section 4.1.1). Droplets produced by the generator must be completely dried at an appropriate rate to ensure that the resulting particles are spherical and non-porous. Particles from both types of generator should be exposed to a radioactive source (*e.g.* Kr-85) to equilibrate the electrostatic charge formed on the particles during generation. This ensures that the calibration aerosol has the same charge distribution from one test to the next, and removes one possible source of variability.

Polymer latex particles are completely characterised by their manufacturers, thus further studies to demonstrate traceability to the international metre is unnecessary. However, the authors recommend carrying out a quality check on the aerosol using a real-time technique during sampling to ensure the detection of any deterioration in the particles arising from handling and storage.

Particles produced by the VOAG must be characterised during use. Although the VOAG operating equation can be used to determine the approximate size of particles produced by the generator, this relationship does not provide traceability to the international metre. Calibrant particles must be sized independently using a traceable technique and a real-time, on-line method is the most convenient. Particles should also be collected and examined by microscopy to check their sphericity.

(d) Sampling of Calibrant Aerosol

The flow rate at which the calibration aerosol is sampled must be measured using a traceable technique (see Section 4.2). The choice of flow rate is dependent on the particular application of the device, and may so differ from that recommended by the manufacturer as to necessitate a recalibration. It should be noted that different sampling flow rates may affect such phenomena as particle bounce and internal losses. The temperature, relative humidity of the sampled gas and the ambient pressure should be recorded using traceable methods to allow for conversion between actual and standard flow rates as appropriate.

The duration of a sampling campaign is best determined by experience: sufficient material should be collected on a stage to give an acceptable 'signal-to-noise' during analysis. It is better to err on the side of longevity because the resulting solutions can be diluted during analysis if necessary. However, care should also be taken not to overload any individual stage of a device, and the manufacturer's operating instructions should be consulted for optimum loadings.

(e) Stage-loading Measurements

The object of a calibration exercise is to characterise the size range within which the collection efficiency of the inertial separator varies from 0 to 100%. An ideal separator would have single step from 0 to 100% efficiency, and the particle size at which this occurred would unambiguously define the performance of the separator. However, this behaviour is never observed in practice because of irregularities in the trajectories of the particles, particle bounce and re-entrainment, which results in curves of collection efficiency versus particle size that are referred to as stage characteristic curves.

When calibrating a cascade impactor with a polydisperse aerosol, the particles that collect on each stage are usually sized by optical or electron microscopy. Particles collected in definite, narrow diameter intervals between d and $(d + \delta d)$ are counted, and a similar analysis is made of the aerosol which penetrates the stage and is collected on a flat surface or a membrane filter. The ratio of the particle number within the size band from d to $(d + \delta d)$ found in the deposit to that in the aerosol penetrating the stage is given by:

$$\frac{\xi(d)}{[1 - \xi(d)]}$$

where ξ is the collection efficiency for that size range. This relatively simple procedure is repeated until the complete characteristic curve of the stage has been determined. However, significant systematic errors can arise because of the following factors:

(i) small particles may be missed (especially true for optical microscopy, where particles less than approximately 0.5 μm geometric diameter are below the lower limit of resolution),
(ii) aggregates may be erroneously counted as single particles,
(iii) the sample may be unrepresentative (mainly applicable to slot impactors, which concentrate the deposit on the collection surfaces immediately beneath the slot-ends (Mercer and Chow (1968) and Chang (1974)).

If the particles are sized by microscopy, geometric (physical) rather than aerodynamic diameters are measured, and the relationship between the two parameters may be difficult to determine because of irregular particle shapes and porosity. A more serious problem may arise if the particles are elongated (*e.g.* fibrous), since the shear flow in the region of the impactor jet tends to orientate them so that their long axes are parallel with the direction of flow and perpendicular to the impaction plate (Fuchs (1964)). Thus, the measured aerodynamic diameter will be greater than the correct value based on random orientation.

Optical particle analysers and the TSI Aerodynamic Particle Sizer have been used to count standard particles during the calibration of impactors and cyclones. However, particles of roughly equal size produce signals in these monitoring devices that spread across neighbouring channels of the multichannel analyser. The likelihood of cross-sensitivity increases as the size range of each channel narrows, making it desirable to group several channels together in an instrument that has more than six to eight channels spanning the size range of the impactor or cyclone. This procedure has the added advantage that statistical errors are reduced by increasing the count for each size band. On the other hand, reduced resolution can be a problem, particularly in the central region of the efficiency curve where the rate of change of efficiency with particle size is greatest.

Many workers prefer to calibrate aerosol analysers with monodisperse particles, despite the increased effort required. Data analysis and interpretation are much easier when spherical, monodisperse particles are used, since their aerodynamic particle sizes are usually well defined and orientation effects do not occur. Samplers have been calibrated with monodisperse solid and liquid particles; particle bounce is more likely to occur when solid particles are used (Franzen and Fissan (1979)), while liquid droplets may shatter on impact with the collection plates. The recommended approach is to calibrate with solid particles if the sampler is to be used to analyse solid particles, and with liquid droplets when sprays and mists are being studied.

Two methods have been developed to calibrate impactors and cyclones with monodisperse particles (May (1975), Rao and Whitby (1978a and 1978b), Boulaud *et al* (1981), Lee *et al* (1985) and Mitchell *et al* (1988)):

(i) Well-characterised polymer latex particles are used, and the particle number concentration is measured upstream and downstream of the sampler. The procedure is repeated at different particle sizes until the collection efficiency curve has been derived. The size corresponding to a collection efficiency of 50% defines the performance of the sampler and is often referred to as the effective cut-off diameter (ECD). If N' and N are the upstream and downstream particle concentrations respectively, the collection efficiency ($\xi(d_1)$) of a single stage for particles of size d_1 is given by:

$$\xi(d_1) = \left[1 - \frac{N}{N'}\right]$$

The experimental procedure for calibrating cascade samplers has been described in detail by Rao and Whitby (1978b). The optical particle counter is located downstream of the impactor for all the measurements, and the airstream leaving the impactor is split so that any excess air is diverted away from the counter. Only the relevant stages of the impactor need to be assembled for calibration: the stage of interest and earlier stages (Figure 16). The upstream concentration (N') is obtained by making measurements without the collection plate in position beneath the stage being calibrated, and the downstream concentration (N) is measured with this collection plate in position.

(ii) Aerosol analysers can be calibrated with reasonable ease using particles detected by gravimetric or chemical methods. Popular choices of solid calibration particles are methylene blue, ammonium fluorescein and nigrosine dye. Particles collected on the individual stages of a device are recovered for analysis by washing several times with an appropriate solvent to achieve complete recovery of the aerosol deposit. Subsequent washings may be necessary with a solvent in which the collected material exhibits greater solubility. Care must be taken to demonstrate that the recovered particles are completely dispersed within the solvent. This requirement is particularly important in the case of fluorescent polymer latex particles as the dye must be released from the polymer matrix. The resulting solutions are made up to a standard volume and analysed using an appropriate technique to determine the mass of particles collected on each stage. The technique must be calibrated by serial dilution of a standard solution of the material from which the aerosol was generated. Calibrations for each of the solvents or solvent mixtures must be performed. Du Pont Pontamine Fast Turquoise 8GLP dye has been

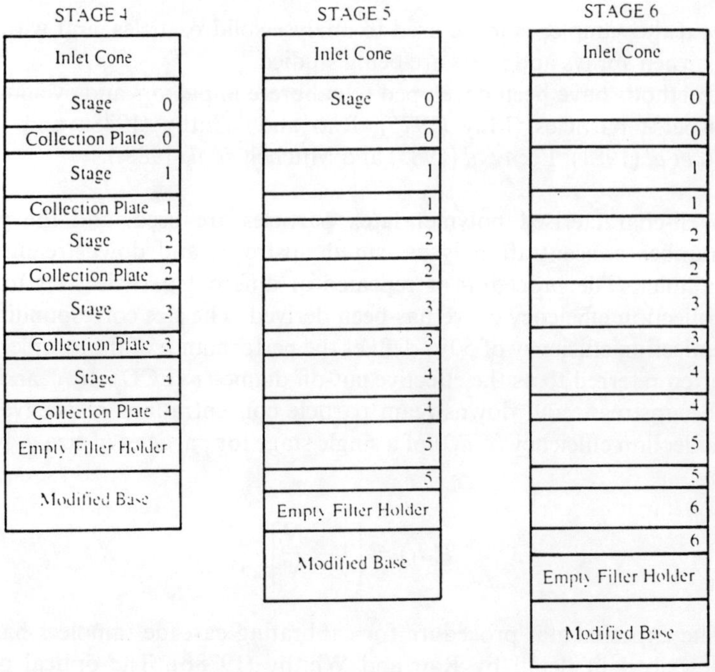

STAGE 4

Inlet Cone
Stage 0
Collection Plate 0
Stage 1
Collection Plate 1
Stage 2
Collection Plate 2
Stage 3
Collection Plate 3
Stage 4
Collection Plate 4
Empty Filter Holder
Modified Base

STAGE 5

Inlet Cone
Stage 0
0
1
1
2
2
3
3
4
4
5
5
Empty Filter Holder
Modified Base

STAGE 6

Inlet Cone
0
0
1
1
2
2
3
3
4
4
5
5
6
6
Empty Filter Holder
Modified Base

Figure 16 *Assembly of cascade impactor for calibration with polymer latex particles*

used for calibrating cyclones at moderate temperatures (177 °C) that would result in the decomposition of other materials (Smith *et al* (1983)). DOP and oleic acid are often chosen as liquid droplet calibrants; these standard aerosols are usually generated by means of a dispersion-type generator such as the VOAG or STAG.

(f) Data Analysis and Reporting

The efficiency of stage i (ξ_i) in a cascade sampler containing X stages (including the filter) is given by:

$$\xi_i = \left[\frac{M_i}{\sum_i^x M_i} \right] \times 100$$

where M_i is the mass of particles collected on the i-th stage. The term $\sum_i^x M_i$ represents the sum of the mass loadings from stage i to the bottom stage or filter. This efficiency is plotted against aerodynamic diameter to produce a collection efficiency curve for each stage. A mathematical function is fitted to the data, and the aerodynamic diameter corresponding to a 50% collection efficiency (d_{50}) is determined for each stage (example data are shown in Figures 17 and 18 for a cascade impactor and cyclone train, respectively).

If a full collection efficiency curve is required, as many data points as possible within the restraints of time and resources should be obtained to cover completely the collection efficiency range from 0 to 100%. These data can be fitted to a tanh function using the Rosenbrock multi-variable non-linear optimisation routine, and the d_{50} value is calculated from the fitted function. This comprehensive approach may be appropriate if, for example, a newly developed instrument is being calibrated for the first time. However, if a more routine calibration check is being performed, fewer data points are necessary. A minimum of three data points which fall on the straight portion of the collection efficiency curve (*e.g.* from 20 to 80%) should be obtained, and at least one of these points should be chosen so that the measured collection efficiency falls close to 50%. These data points can be fitted to a straight line by linear regression, and the individual d_{50} values calculated from the fitted functions.

Virtual impactors do not have particle collection surfaces and both fractions of the aerosol remain airborne after inertial separation has occurred. The calibration of virtual impactors has been described by Loo *et al* (1976), O'Connor (1973) and Chen *et al* (1985b). Polystyrene latex particles can be used, and particles in both the minor and major flows are counted alternatively with an Aerodynamic Particle Sizer. The collection efficiency is determined by:

$$\xi = \frac{C_a}{C_a + \left[\dfrac{C_b Q_b}{Q_b - Q_c}\right]}$$

where C_a and C_b are the number concentrations measured in the minor and major outlet flows respectively, and Q_b and Q_c are the volumetric flow rates in the major and bleed-off airstreams respectively. At low flow rates, the bleed-off flow is replaced with dilution air to satisfy the requirements of the particle counter, and the collection efficiency of the virtual impactor is calculated according to:

$$\xi = \frac{C_a}{C_a + C_b}$$

Wall losses have been measured by Chen *et al* (1985b), and a virtual impactor calibrated using a VOAG and liquid droplet–fluorescent tracer technique. Both the major and minor flows from the impactor were drawn through separate filters to remove the DOP for analysis. The collection efficiency (ξ) was calculated for each particle size using the equation:

$$\xi = \frac{m}{m + M}$$

where m is the mass of fluorescein collected from the minor flow (large

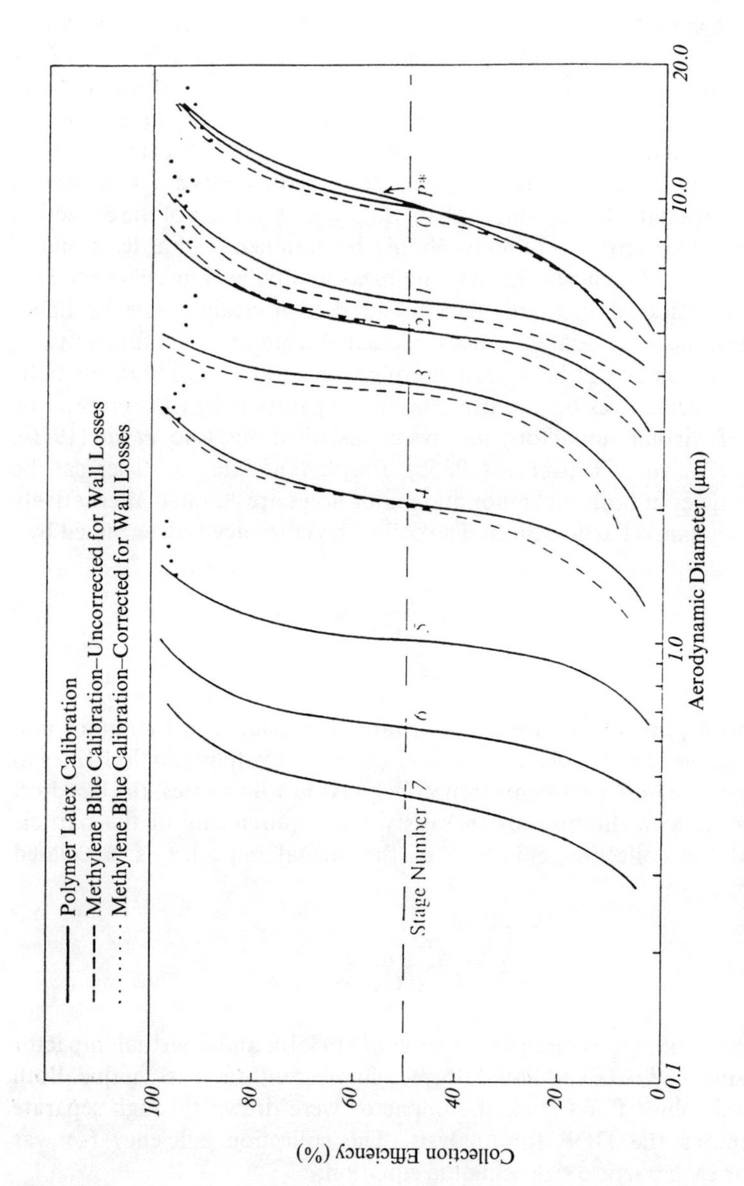

Figure 17 *Cascade impactor calibration curves*

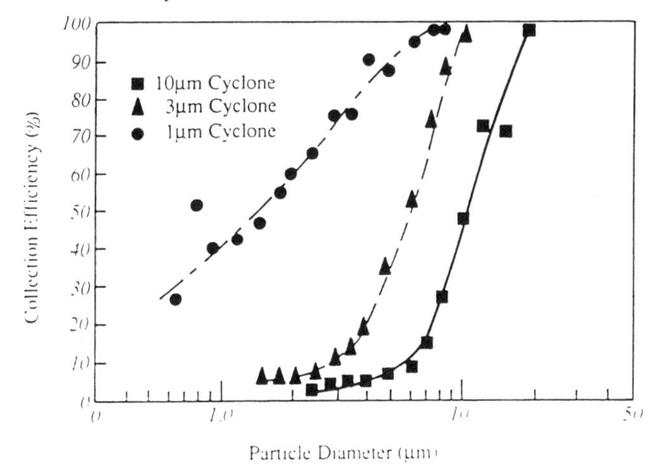

Figure 18 *Calibration curves for three cyclones*
(Reprinted by permission of Sandia National Laboratories from 'Particle Collection by Cyclones at High Temperature and Pressure', by J. C. F. Wang and M. A. Libkind, SAND 82–8611. Copyright 1990 by Sandia National Laboratories, Albuquerque, New Mexico 87185, USA.)

particles), and M is the mass of fluorescein deposited from the major flow (small particles). Wall losses were obtained by washing the inside surfaces of the impactor, and applying the expression:

$$\mathrm{WL}_i = \frac{M_{\mathrm{w}_i}}{M_{\mathrm{w}} + m + M}$$

where WL_i is the wall loss on the i-th section of the impactor, and $M_{\mathrm{w}i}$ and M_{w} are the mass of fluorescein deposited on the i-th section and the total mass collected on all internal surfaces, respectively.

(g) Data Anomalies and Distortions

Collection efficiency curves can be distorted by non-ideal behaviour arising from particle bounce (curve reaches a plateau at an efficiency of less than 100% and usually decreases at larger sizes), particle filtration when glass fibre or paper filters are used as collection surfaces (curve rises less steeply), and surface irregularities (collection efficiency increases at small particle sizes to produce a prolonged tail in the characteristic curve). Particle bounce may cause a significant shift in the ECD for a given stage. Wall losses should always be expected in impactors when particles are being measured with aerodynamic diameters exceeding 5 μm. Such behaviour cannot be detected from the shape of this calibration curve and must be measured by mass balance studies of the aerosol deposited throughout the sampler, including the internal walls (O'Connor (1973)).

Care is needed to ensure that electrostatic charges associated with the calibration particles are minimised, since particles produced by atomisation methods are usually highly charged. Charge equilibration is usually carried out by passing the calibrant aerosol through a tube containing a Kr-85 radioactive source; the resulting particles emerge with a low charge corresponding to the Boltzmann distribution. Horton and Mitchell (1990 and 1992) reported the calibration of a California Measurements QCM low-pressure, cascade impactor with methylene blue particles produced by a VOAG: charged 5 μm diameter particles deposited some way from the sensitive region of the oscillating quartz crystal to give erroneous results, while similar particles deposited correctly in the central region of the sensing crystal after passage through a charge equilibrator.

5.1.2 Inertial Spectrometers

All the particles that pass through an inertial spectrometer are collected on a single deposition surface (metal foil or membrane filter) after size separation has occurred, and the purpose of calibration is to establish the position on the surface at which particles of a given aerodynamic diameter deposit. Their precise location is normally measured with respect to the inlet to the spectrometer. The entire deposit can be scanned by a microscope, and the aerodynamic size of the particles can be determined as a function of deposition distance from the inlet to the spectrometer. Furthermore, the collection medium can be sectioned into smaller size fractions after use for higher resolution of the resulting mass and chemical analysis.

Several instruments have been designed since 1955 that can be classified as inertial spectrometers, including the Goetz (Goetz *et al* (1960)) and Stöber (Stöber and Flaschbart (1969 and 1971), and Stöber *et al* (1972)) centrifuges, and equivalent devices developed at Los Alamos (Tillery (1974)) and the Lovelace Foundation (Kotrappa and Light (1972)), and the Inspec (Prodi *et al* (1979 and 1982)). Spectrometer performance is highly dependent upon operating conditions; separate calibrations must be performed for each set of operating conditions used, and careful control of the operating parameters is required with this type of instrument if high size resolution is required. They are best calibrated with traceable monodisperse polymer latex particles (see Section 4.1.1), since additional useful information can also be obtained from the deposition profiles of the multiplets. Stöber and Flaschbart (1971) calibrated their spiral duct centrifuge with latex microspheres, and were able to resolve aggregates consisting of twenty-three 1.8 μm diameter polystyrene particles from those containing twenty-two particles. Similar studies have been undertaken by Mitchell and Nichols (1988) to calibrate the Inspec with 2 μm diameter polyvinyltoluene microspheres (Figure 19). They showed that this instrument could resolve multiplet aggregates containing as many as eight particles.

It is essential that the spectrometer is leak-tight, and this can be confirmed by blocking the inlet and checking that all flows fall to zero. The location of PSL particles of a given size on the deposition surface is determined by microscopy and recorded, and this procedure is repeated with further particle sizes until the

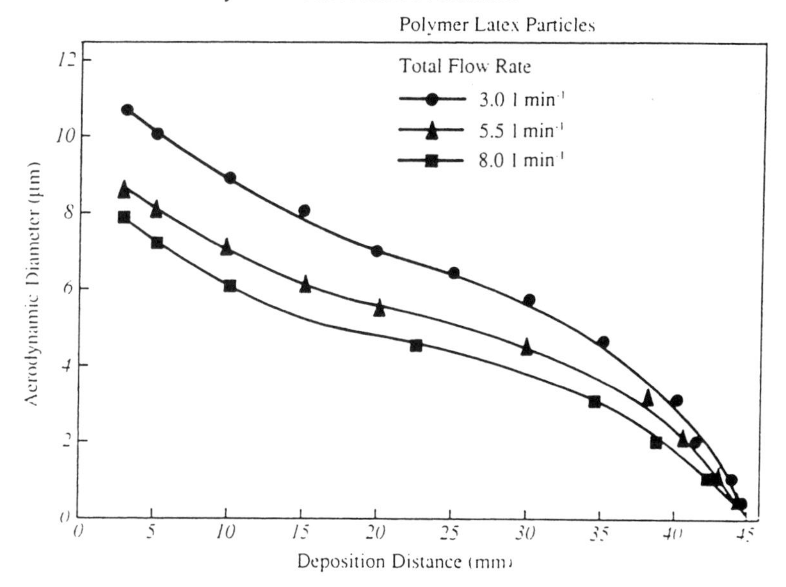

Figure 19 *Calibration curves for Prodi inertial spectrometer*

operating range of the instrument has been covered. A calibration curve is then constructed as a plot of aerodynamic diameter against the distance along the collection medium. This curve is a function of the air flow rates through the spectrometer, and may also depend on other factors specific to the instrument such as the revolution rate of the rotor in a centrifuge. It is essential that the operating parameters of the spectrometer are measured using traceable techniques and duplicated for each of the calibration measurements. This is best achieved by recording the temperature, humidity and ambient pressure during the measurements, and adjusting subsequent flow rates to the same standard flow rate used in the first calibration measurement. Smith (1982) and Mitchell *et al* (1984) calibrated a Stöber spiral duct centrifuge using ten different sizes of latex particles (Figure 20), which is probably the minimum required to obtain good resolution.

Calibration with a polydisperse aerosol is also possible, which gives the added advantage of performing the calibration in one measurement and removing the need to duplicate conditions in subsequent exercises. The deposit is inspected at regular intervals along the length of the substrate, and the size of individual particles located at each position is measured by microscopy-image analysis. A calibration curve is constructed from these data. However, it is more difficult to provide traceability to the international metre using this method as the density and sphericity of all the particles must be known unambiguously to allow the conversion of microscopy-measured diameters to aerodynamic diameters.

Smith (1982), Mitchell *et al* (1984), Martonen (1977) and Stöber *et al* (1978) have reported non-ideal collection behaviour within inertial spectrometers used to sample highly concentrated aerosols. Cloud settling can occur which does not

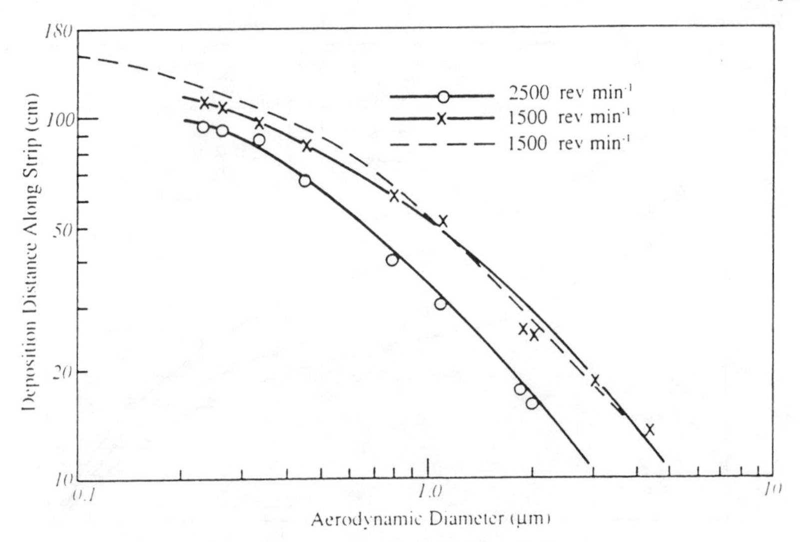

Figure 20 *Calibration curves for Stöber spiral duct centrifuge*

result in a monotonic relationship between particle diameter and distance of the deposit along the collection surface. The onset of this behaviour depends on the fluid dynamics of the spectrometer and cannot be detected when using a monodisperse aerosol.

5.1.3 Real-Time Aerodynamic Particle Sizers

Real-time aerodynamic particle sizers separate particles on the basis of their inertia by accelerating the particles through a well-defined flow field and measuring particle time-of-flight across a split laser beam. Particle time-of-flight is subsequently converted to aerodynamic diameter through an internal function derived from calibration with particles of known diameter. The guidance given below is restricted to the most commonly available versions of this type of instrument, (*i.e.* the TSI Aerodynamic Particle Sizer (APS33B) (superseded recently by the APS3320 to measure the aerodynamic diameter and scattered light intensity of each particle) and the API Aerosizer (API)). Both the APS33 systems and the API measure aerodynamic diameter without collecting or trapping the aerosol particles for quantification (Figure 21).

(a) TSI Aerodynamic Particle Sizer (APS)

Calibration of the APS involves the re-definition of the internally stored function relating aerodynamic diameter to particle time-of-flight. This procedure is performed through the instrument operating software and is compre-

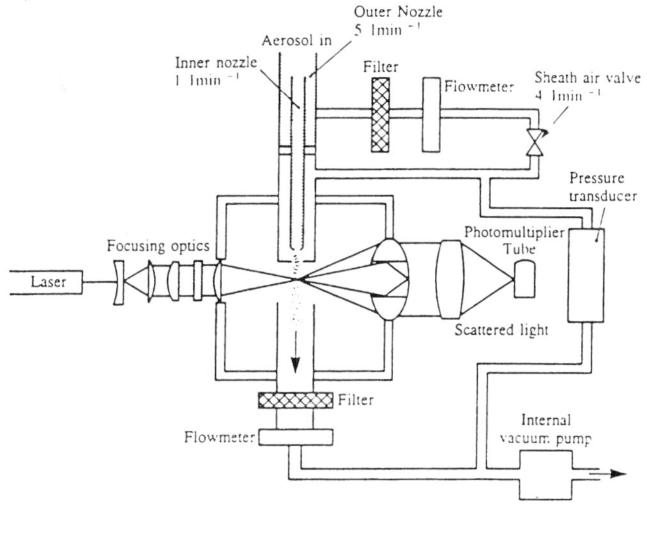

(a) TSI APS33B

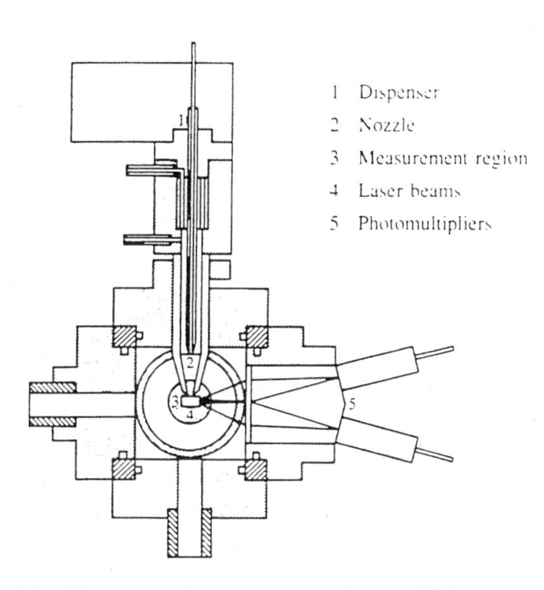

(b) Malvern - Amherst Aerosizer (API)

Figure 21 *Real-time aerodynamic particle sizers*

hensively documented in the manufacturer's operating manual. It is recommended that this procedure be performed annually along with a routine maintenance check by the manufacturer or one of their registered agents. Calibration exercises with the APS33B (Agarwal *et al* (1982), Baron (1984 and 1986), Chen *et al* (1985a) and Griffiths *et al* (1986a and 1986b)) have demon-

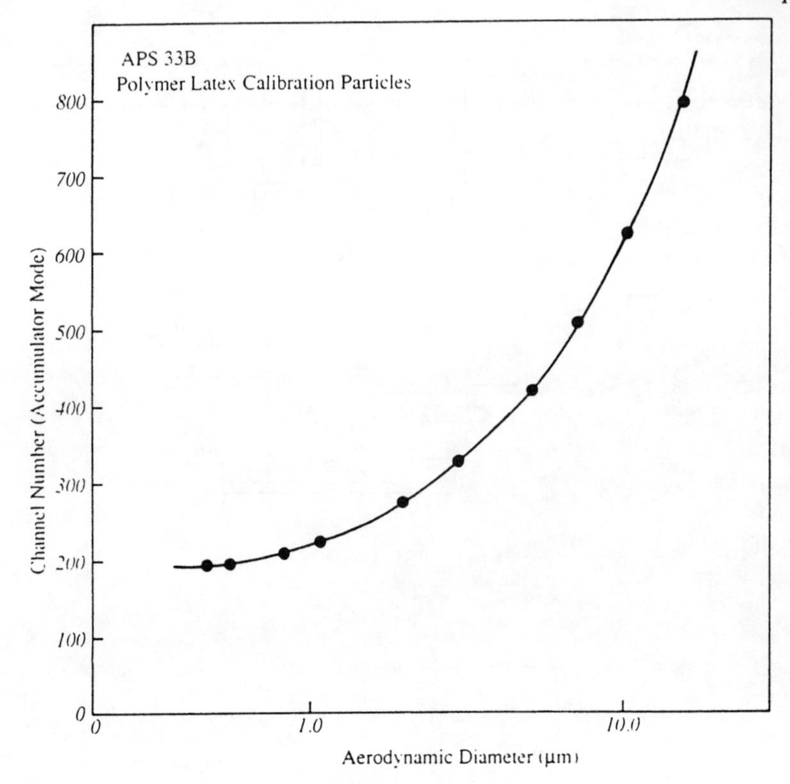

Figure 22 *Calibration curve for TSI aerodynamic particle sizer*

strated the good size resolution of this type of instrument (Figure 22) and have shown that the use of the highest quality monodisperse aerosols is advantageous. Calibrant aerosols should be monodisperse, although a cocktail of monodisperse particles can be used to check the calibration in one measurement. The airborne concentration of the calibrant aerosol should be sufficiently low to prevent problems with particle coincidence within the instrument measurement zone, and the user should refer to the manufacturer's operating instructions for guidance on the limits of airborne concentration to be sampled. Considerable care must be taken to ensure that the operating parameters (flow rates and pressures) are correctly adjusted before calibrating the instrument, because any small variations can have a significant effect on the resulting data (Griffiths *et al* (1986a)). A satisfactory calibration check has been performed if acceptable agreement is reached between the APS-measured diameters and the diameters of the calibrant particles.

For particles of the same aerodynamic diameter, Wilson and Liu (1980) calculated that the particle with the higher density will be defined as being larger by the APS33B. This effect was confirmed experimentally by Baron (1986), who measured an increase of about 7 to 9% for spherical particles with a density of 2 $g\,cm^{-3}$ compared with spherical particles of equivalent size and a density of

1 g cm^{-3}. Baron has also suggested that coincidence may occur between sub-micron particles not normally detected by the APS33B, and between any electronic noise and larger particles. Blackford *et al* (1988) have reported that particles smaller than 0.7 μm diameter may not be detected at 100% efficiency; Kinney and Pui (1985) have measured inlet efficiences of 90% for 3 μm diameter particles to less than 45% for particles larger than 10 μm diameter when sampling from a low-flow system (an alternative inlet design was also proposed for the measurement of particle size distributions).

It should be noted that the APS can distort liquid droplets during measurement and this effect invalidates any calibration with solid particles (Baron (1986) and Griffiths *et al* (1986b)). Liquid droplets are distorted to form oblate spheroids in the high-velocity flow field within the region of the laser beams, and appear smaller in aerodynamic diameter than they would be if they were spherical. The calibration curve can be displaced by as much as 20% for droplets of ~15 μm diameter (Figure 23), but this effect decreases as the droplets are reduced in size to become insignificant for droplets smaller than 2 μm diameter. If the instrument is to be routinely used to measure a liquid-droplet aerosol, a complete recalibration with traceable droplets of the same

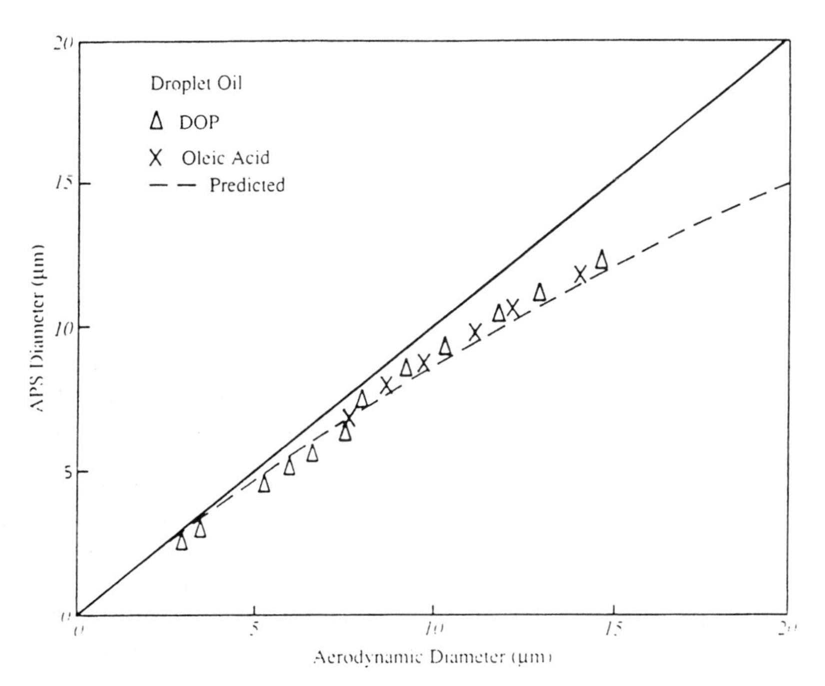

Figure 23 *Effect of liquid droplet distortion on the calibration of a TSI aerodynamic particle sizer*
(Reprinted by permission of Elsevier Science Ltd. from 'Calibration and Use of the Aerodynamic Particle Sizer (APS 3300)', by P. A. Baron, *Aerosol Sci. Technol.*, **5**, 55. Copyright 1986 by American Association for Aerosol Research.)

liquid is necessary because the degree of distortion is dependent on the viscosity and surface tension of the particular liquid. In addition, the APS is known to underestimate the size of non-spherical particles and cannot be used to make traceable measurements on such particles.

A user should carry out a routine calibration check before using the APS to guarantee good quality measurements. This check can be limited to determining the ability of the APS to size correctly a minimum of three discrete diameters of traceable particles over the size range of interest. The APS should be operated at the flow rates and pressure drop listed on the instrument data sheet which accompanies the record of the previous recalibration, since the internal calibration function of the instrument is not valid unless these conditions are duplicated. The operating manual describes a procedure to set these parameters, and lists their acceptable tolerances based on the instrument sampling dry, particle-free air. It may prove difficult to achieve these settings if the instrument has been used to sample high-concentration aerosols that have started to clog the filters protecting the internal flow meters; these filters require replacement before continuing the calibration exercise. The density of the particles must also be known and entered in the operating software in order to correct the data and take into account the difference between the measured particle density and that of the calibration particles (Wang and John (1987)). Thus, it is assumed that all of the particles have the same uniform density for this correction to be valid.

[Authors' note: The Model 3320 Aerodynamic Particle Sizer has recently superseded previous models (3300 and 3310). A double-crest, time-of-flight optical system creates only one signal as an aerosol particle passes through the measurement zone. This approach eliminates the need to correlate two separate pulses, and reduces coincidence counts. However, the new system requires independent testing and evaluation in the laboratory and under representative field conditions.]

(b) API Aerosizer (API)

A complete recalibration of the API is impossible to perform, and it is recommended that routine annual maintenance be carried out by the manufacturer or their registered agents. The flow rate through the measurement volume is sonic and the internal calibration relies upon this condition being achieved. If the flow rate does not reach sonic velocities due to defects in the flow circuit, the operating software will not permit the instrument to be used until the fault is rectified. No facility exists to set instrument flow rates and operating pressures. As with the APS, the API cannot provide traceable measurements of non-spherical particles, and furthermore cannot be recalibrated to measure liquid droplets.

A routine calibration check must be performed before using the API to make good quality measurements, and an identical procedure to that described above for the APS should be adopted. The comments concerning particle density are equally applicable to the API. Although the airborne concentration for a given particle size can be greater with the API than the APS before coincidence problems are encountered, an upper limit does exist for the API and the manufacturer's operating instructions should be consulted.

Both the APS33B and API have been shown to undersize solid non-spherical particle shape standards (Figure 24). The effect is substantial: 15 μm aerodynamic diameter particles may be undersized by as much as 25% in the APS33B and 40% in the API (Marshall *et al* (1991) and Marshall and Mitchell (1991)). Theoretical studies indicate that the dynamic shape factor of the calibrant particles increases as they are accelerated, although further work with a wide variety of different particle shapes is required to confirm this hypothesis.

5.2 Sedimentation/Elutriation Techniques

Sedimentation techniques form a single class of aerosol analysers that determine directly particle aerodynamic diameter, or rather the particle Stokes diameter. Sedimentation devices are encountered in a limited number of specialised applications, such as occupational hygiene in which low aerosol flow rates are experienced.

5.2.1 Sedimentation Cells

Sedimentation cells rely on particle settling under gravity to measure particle diameter (Stahlhofen *et al* (1975)). The aerosol is introduced into a small volume (typically less than 1 cm^3), which is subsequently sealed to ensure that the particles settle through still air. Aerosol particles are illuminated through an optical window by an intense beam of light and are viewed through a horizontal microscope in a direction perpendicular to the light beam. The time taken for particles to settle through a well-defined distance between two lines of reference is recorded, and used to determine the particle settling velocity and diameter. Sedimentation cells are only appropriate for the measurement of aerodynamic diameters in the range from 0.3 to 5 μm because of the following:

(a) Brownian motion of particles smaller than 0.3 μm introduces unacceptably large variations in the measured settling velocity; such particles also have a tendency to drift out of the field of view during the measurement.
(b) Particles larger than 5 μm diameter settle too quickly for accurate time measurements to be made.

The heating effect of the light beam will induce thermal convection within the chamber, and must be minimised. While the magnification of the microscope is chosen so that the smallest particles encountered are sufficiently well illuminated to be easily observed, care must be taken to eliminate any operator bias toward the selection of the larger and brighter particles when measuring a polydisperse aerosol. Operator performance can be assessed by measuring solid spherical particles of know diameter and density.

The main sources of error encounted with a sedimentation cell are the uncertainty in the distance between the reference lines, a non-representative

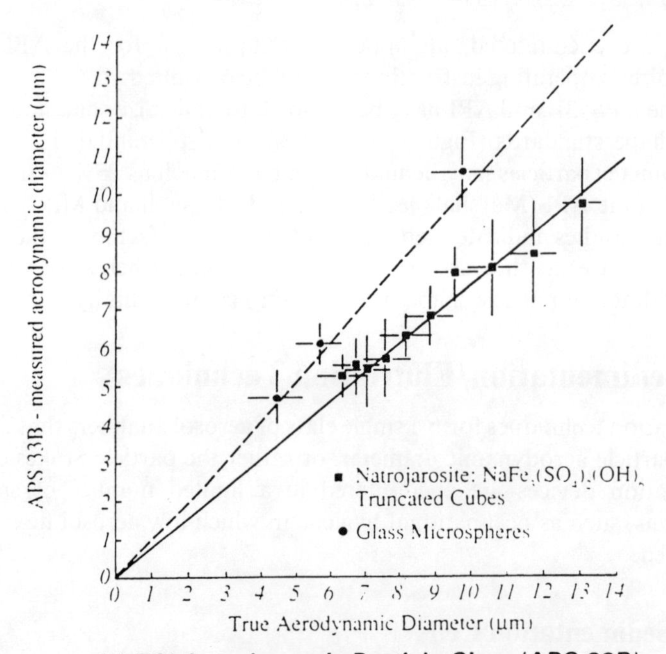

(a) TSI Aerodynamic Particle Sizer (APS 33B)

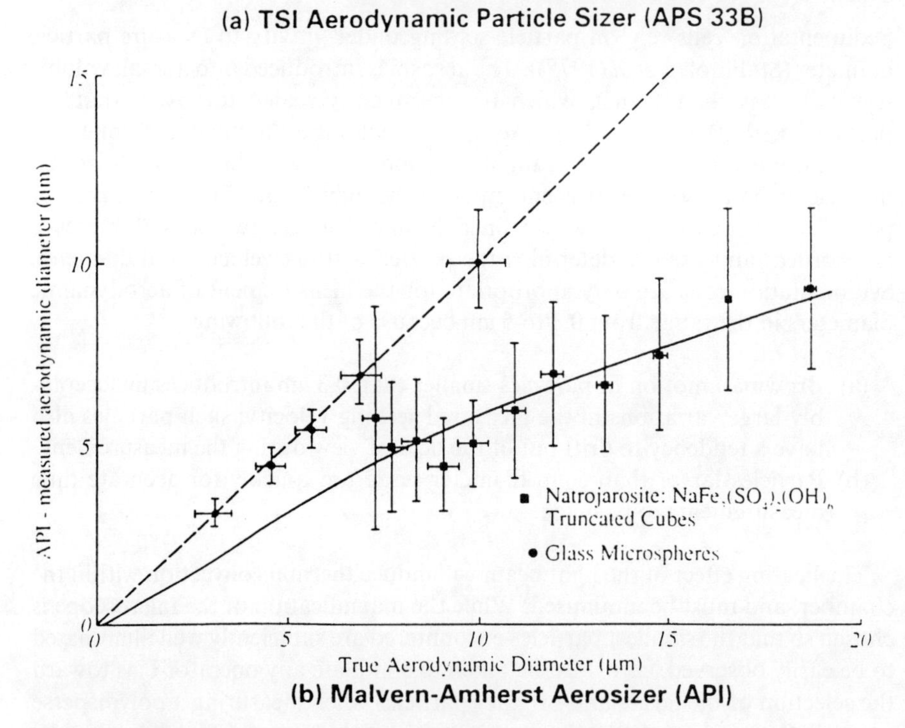

(b) Malvern-Amherst Aerosizer (API)

Figure 24 *Influence of particle shape on APS33B and API responses*

sample within the cell, the measurement of too small a number of particles, the precision with which the operator makes the time measurement, and the biasing tendency of an operator to select larger particles. The only form of calibration required for such devices is the determination of the distance between the reference lines and a calibration of the timing device.

5.2.2 Elutriators

Elutriators are another class of instrument which rely on particle settling through a fluid flow to determine their diameters. Such devices can be either vertical or horizontal in orientation, although vertical elutriators are not normally used to measure particle diameters and are therefore not covered in this document. Two flow streams are used: a clean air stream and the aerosol stream. A horizontal laminar flow of clean air is established in a horizontal duct of rectangular cross-section. Aerosol is introduced as a thin stream along the upper surface of the duct and particle settling occurs perpendicular to the gas streamlines. Particles of different size settle at different rates and deposit at different locations along the duct floor. It is common to use a duct width that increases along its length to reduce the clean gas flow velocity and permit smaller particles to deposit at convenient distances from the entry point. Gravitational settling of the particles is a relatively slow process, and the clean air flow must be at a low velocity so that the duct is of manageable length; under such circumstances, these devices are sensitive to convection. The floor of the duct is lined with glass slides or a metal foil that can be removed in sections and analysed for the number or mass of deposited particles. Each section corresponds to a range of terminal settling velocities that in turn defines a range of particle diameters.

The particles are size-separated under the influence of gravity according to Stokes Law. The terminal velocity (v_t) of a particle of volume equivalent diameter d_{ve} is given by:

$$v_t = \frac{(\rho_p - \rho_g) d_{ve}^2 g}{18 \mu \chi}$$

where ρ_p and ρ_g are the particle and gas density respectively, μ is the gas viscosity, g is the acceleration due to gravity, and χ is the particle dynamic shape factor (unity for spheres). For particles in a gaseous medium, $\rho_p \gg \rho_g$ and the latter can be neglected. This equation is only valid if the particle motion is unaffected by wall interactions, terminal velocities have been reached, and particle–particle interactions are negligible. The first two criteria are easy to satisfy with aerosol-based systems, but the third requirement limits the particle concentration to below 100 mg m^{-3}.

Elutriators are occasionally used to determine aerodynamic diameter (d_{ae}) directly from a calibration of deposition distance versus particle size, using spherical particles of known density (Griffiths *et al* (1984) and Marshall *et al*

(1990)). The Timbrell spectrometer (Timbrell (1972)) can be used to achieve a high degree of size-grading of aerosols containing particles in the range from about 2 to 20 μm aerodynamic diameter. Particles are winnowed by a recirculating flow of clean air in a horizontal settling chamber, and the size-separated particles are collected on a series of microscope slides located along the bottom of the chamber (Figure 25). Griffiths *et al* (1984) reported the use of polydisperse, near-unit density, spherical particles of commercial hairspray (polyvinyl acetate) as a useful calibrant for the Timbrell spectrometer. Marshall *et al* (1990) described a calibration procedure using polystyrene latex and polyvinyl acetate microspheres: optimum size resolution (close to 12%) was achieved in the range from 4 to 12 μm aerodynamic diameter, when the incoming aerosol flow rate was 1 $cm^3\ min^{-1}$ and the ratio of winnowing to aerosol flow rates was 150:1 (these data are shown in Figure 26).

There are many designs of sedimentation device, including the elutriator of McCormack and Hilliard (1980), which provided aerodynamic size distributions of sodium fire aerosols that were in good agreement with aerodynamic size distribution data from an Andersen Mark-II cascade impactor. The horizontal sedimentation battery of Boulaud *et al* (1983) is a development of the basic sedimentometer; incoming aerosol is introduced into 10 rectangular channels via a 30 cm long cone to ensure that the aerosol is homogeneous (Figure 27). This instrument provides reliable mass–aerodynamic size distribution data over the aerodynamic diameter range from 4 to 9 μm. The Infrasizer Mk III is a laminar flow particle classifier developed by Raimondo *et al* (1979) to separate and size particles of a few to hundreds of microns.

5.3 Optical Particle Techniques

A large group of aerosol analysers operate on the basis of light interaction with particles. These instruments may be divided into five categories (see Table 2.1):

(a) classical and active-cavity laser light scattering instruments (optical particle counters),
(b) laser (Fraunhofer/Mie) diffraction analysers,
(c) laser–Doppler techniques,
(d) light intensity deconvolution analysers,
(e) GALAI-CIS time-of-interaction analyser.

The first four systems are widely used, while the potential of the time-of-interaction analyser has yet to be fully exploited for aerosol analysis. Single-particle light scattering is used to resolve particles in the size range 0.1 to 100 μm diameter. Laser diffractometers are field-scattering systems that measure the size distribution of a population of particles simultaneously, and have a wider operating range (0.1 to 1000 μm diameter); they also have the added advantage that they can be operated in a non-invasive manner. Laser phase–Doppler systems are generally more expensive than diffractometers, are non-invasive and can provide additional information on particle velocity. Test aerosols of

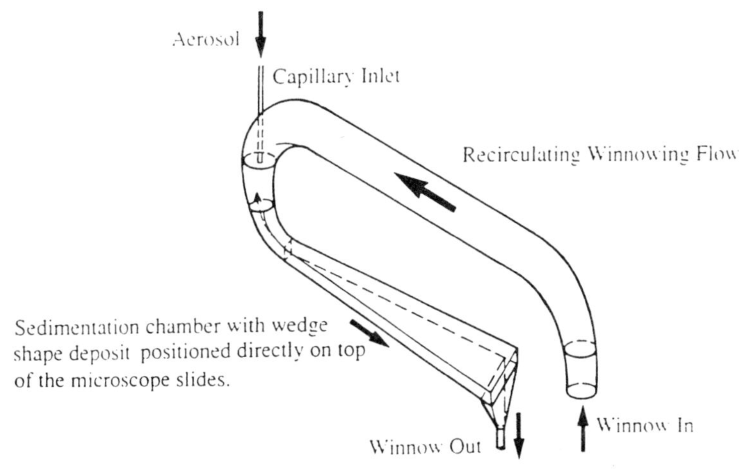

(a) **Timbrell Aerosol Spectrometer**

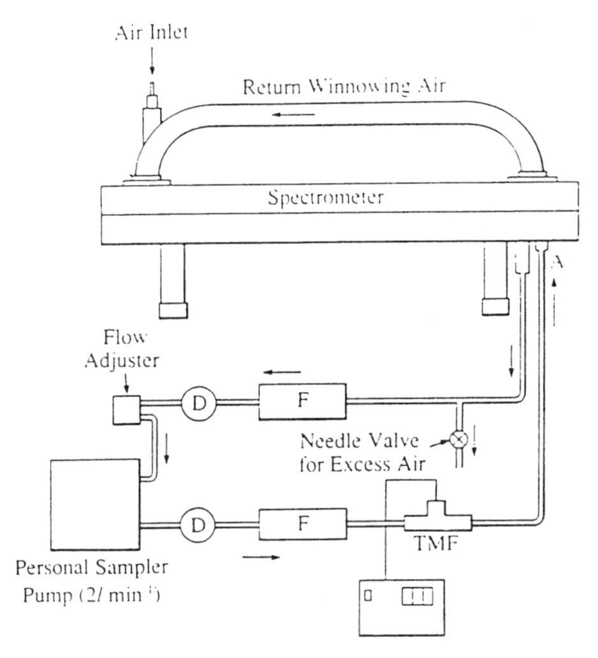

(b) **Flow Circuit**

Figure 25 *Timbrell sedimentation aerosol spectrometer*

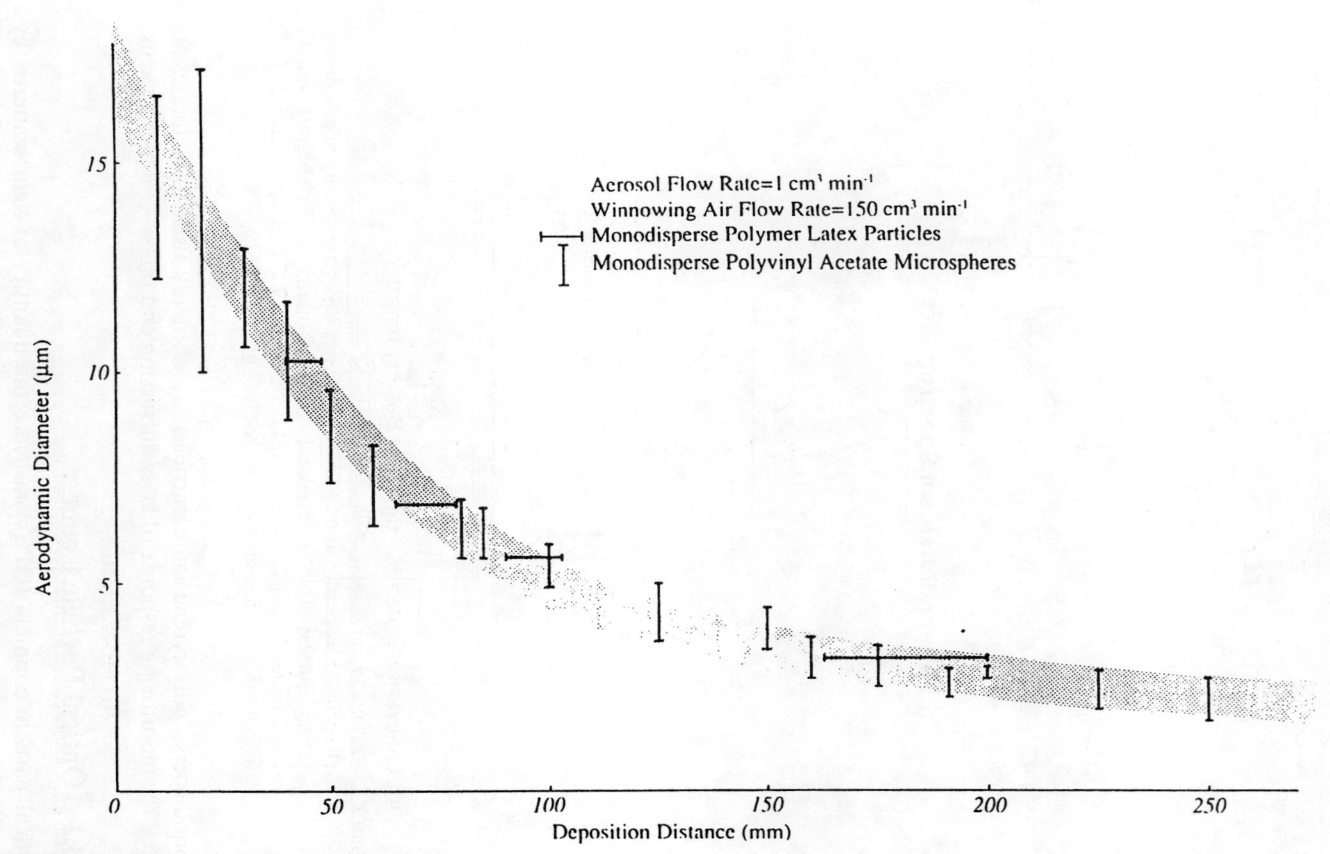

Figure 26 *Calibration of a Timbrell spectrometer*

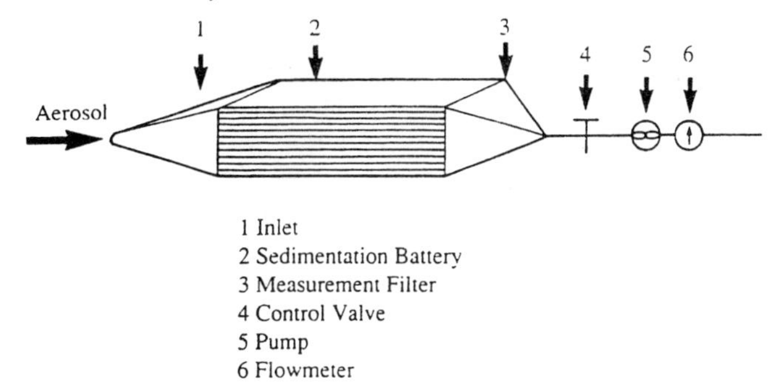

(a) Schematic Diagram of the Sedimentation Battery and Aerosol Flow Arrangement

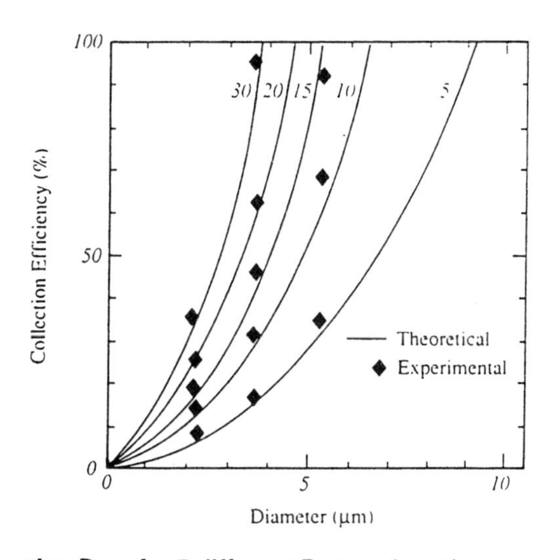

(b) Calibration Data for 5 different Battery lengths

Figure 27 *Sedimentation battery of Boulaud* et al
(Reprinted by permission of Elsevier Science Ltd. from 'Study of a Sedimentation Battery', by D. Boulaud, J. C. Chouard, C. Frambourt and G. Madelaine, *J. Aerosol Sci.*, **14**, 421. Copyright 1983 by Elsevier Science Ltd., The Boulevard, Langford Lane, Kidlington, OX5 1GB, UK.)

monodisperse polystyrene latex particles are normally used to calibrate and check the monitoring response of laser diffraction and phase–Doppler systems.

5.3.1 Optical Particle Counters

Optical particle counters (OPCs) measure the quantity of light scattered by individual particles as they pass through a beam of intense monochromatic or white light, which distinguishes them from instruments based upon Fraunhofer diffraction (Malvern-type particle sizers) and laser (phase) Doppler velocimetry. A beam of light is focused onto a 'view-volume' through which the particles pass and are individually detected (Figure 28). The amount of scattered light is measured by a photosensitive detector, and the magnitude of the electrical signal fed to a multichannel analyser is proportional to the particle size. Several types of OPC are commercially available and their particle size limits vary from about 0.1 μm for open-cavity laser-based instruments of the Knollenberg-type (Pinnick and Auvermann (1979)) to 0.3 to more than 50 μm for the Polytec (Schegk *et al* (1984) and Marshall *et al* (1988a)). The number of accumulating channels varies from four to five in some instruments designed for clean-room monitoring, and can be as high as 128 in some counters.

The objective of calibrating an OPC is to determine the relationship between the intensity of scattered light and particle size. This relationship can be calculated by applying Mie theory (Mie (1908) and Kerker (1969)), which agrees reasonably well with experimental data for spherical particles of known refractive index. However, this theoretical approach is often unsuitable for more complex shaped particles, and experimental calibrations are necessary for most applications.

Polymer latex particles (see Section 4.1.1) are used by almost all workers as calibration standards for OPCs. However, the refractive index of the calibrant material can affect the instrument sensitivity, as reported for example by Szymanski and Liu (1986) for the Knollenberg-type OPC (Figure 29). Similar observations have also been made with other types of OPC (Whitby and Vomela (1967), Willeke and Liu (1976), Pinnick *et al* (1973) and Liu *et al* (1974)), and it is generally accepted that accurate work merits calibration of the OPC with independently sized particles of the material to be studied. This self-consistent approach is seldom feasible since most aerosols of interest consist of ill-defined polydisperse particles of irregular shape. Therefore, a calibration procedure for OPCs has been developed which involves the use of an inertial pre-collector of known ECD (effective cut-off diameter), as described by Marple and Rubow (1976) and Marple (1979) and carried out by Fissan and Helsper (1983), Buettner (1985) and Marshall *et al* (1988b). Single-stage impactors and cyclones are used as the pre-collector; thus the OPC is calibrated with respect to the aerodynamic sizing characteristics of the pre-collector rather than to diameters related to the light scattering cross-section of the calibrant. Such an approach is advantageous because aerodynamic measurements are closely related to the transport properties of particles throughout the size range of most OPCs, whereas their optical properties are of no significance in defining this behaviour.

The calibration of an OPC with a pre-collector depends on generating a polydisperse test aerosol with a size distribution that matches the collection efficiency–particle size curve of the pre-collector. This particle size distribution is measured by the OPC without the pre-collector, and the process is repeated with the pre-collector placed immediately in front of the OPC. If the OPC has i channels, the number count in each channel from the first measurement can be represented by n_1^1, n_2^1, n_3^1...... n_i^1, and for the second

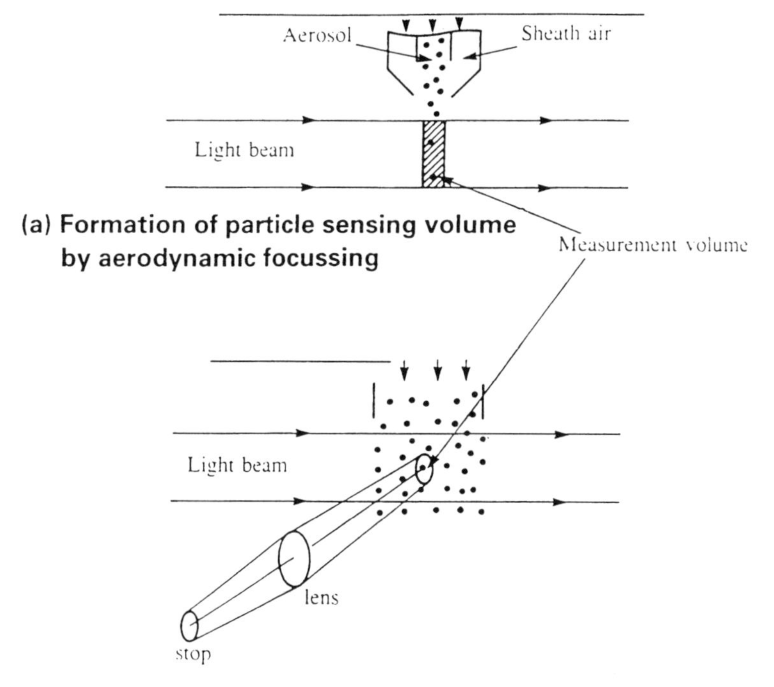

(a) Formation of particle sensing volume by aerodynamic focussing

(b) Formation of particle sensing volume by optical means

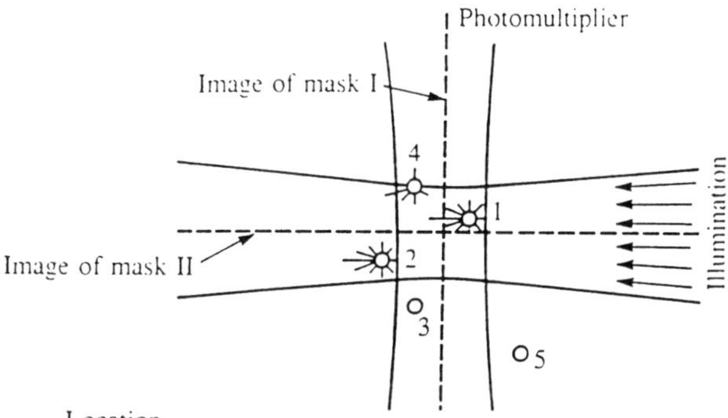

Location

1 Particle illuminated fully in full view of detector - sized correctly

2 Particle illuminated fully out of view of detector - not measured

3 Particle in full view of detector but unilluminated - not measured

4 Particle partly illuminated in full view of detector - undersized (border-zone error)

5 Particle unilluminated and out of view of detector - not measured

(c) Origin of border zone error

Figure 28 *Optical particle counters—detection methods*

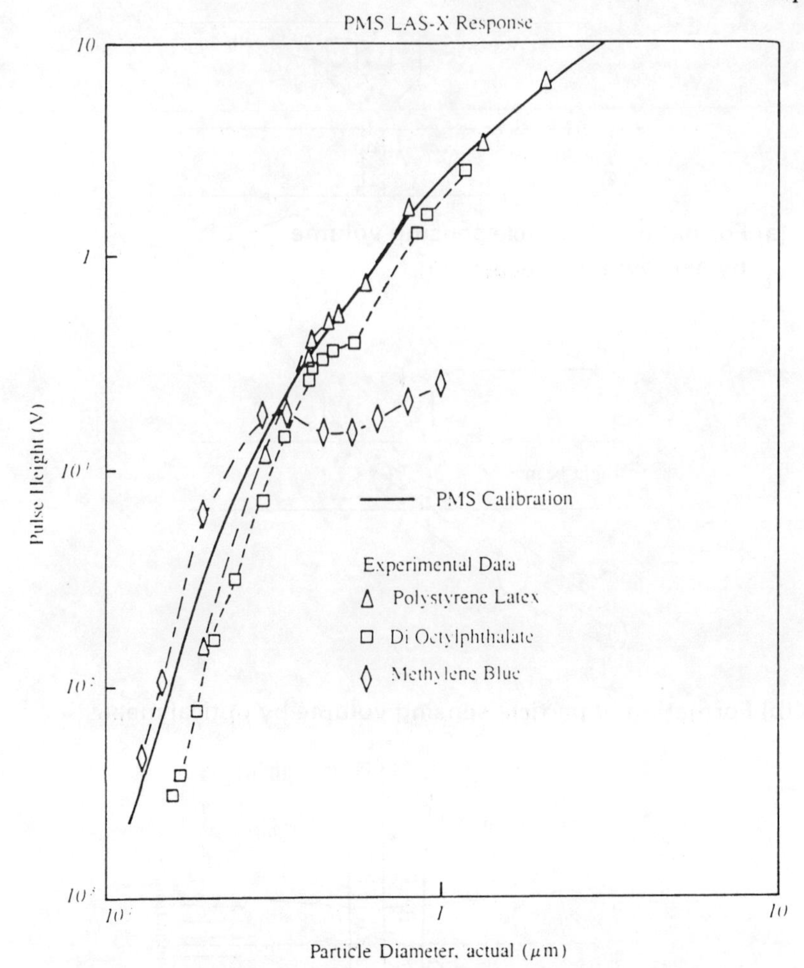

Figure 29 *Response of a Knollenberg-type OPC to particles with different refractive indices*
(Reprinted by permission of B. Y. H. Liu and VCH Verlagsgesellschaft MbH from 'On the Sizing Accuracy of Laser Optical Particle Counters', by W. W. Szymanski and B. Y. H. Liu, *Part. Charact.*, **3**, 1. Copyright 1986 by VCH Verlagsgesellschaft MbH, Postfach 101161, D-69451 Weinheim, Germany.)

measurement by $n_1^2, n_2^2, n_3^2 \ldots\ldots n_i^2$. The attenuation of the particle concentration by the pre-collector (A) for the x-th channel of the OPC is given by:

$$A_x = \left[1 - \frac{n_x^2}{n_x^1}\right]$$

If the pre-collector and aerosol characteristics are chosen correctly, this attenuation varies from 0% in the lowest channels of the OPC and rises in a smooth curve to 100% in the higher channels. The channel number in which A is closest to 50% corresponds to the ECD of the pre-collector, providing a direct link between this channel number of the OPC and the aerodynamic diameter. If the collection efficiency curve of the pre-collector is known, other calibration points can be determined by matching values of A with the corresponding collection efficiencies of the pre-collector. However, it is more usual to repeat the whole procedure several times with pre-collectors that have different ECD values, because they can be more accurately defined than a complete efficiency curve (Buettner (1985)). The calibration of an OPC by this method is rapid and easy to perform, but the accuracy depends on the number concentration and size distribution of the test aerosol remaining constant for each set of measurements (Figure 30).

Since OPCs detect individual particles, they are frequently used to measure number concentrations. These measurements are subject to errors caused by particle coincidence, but generally they are insignificant unless the aerosol concentration exceeds about 1000 particles cm^{-3}. The calibration of OPCs against an aerosol of known number concentration is usually carried out by comparing the total number concentration measured by the OPC with the

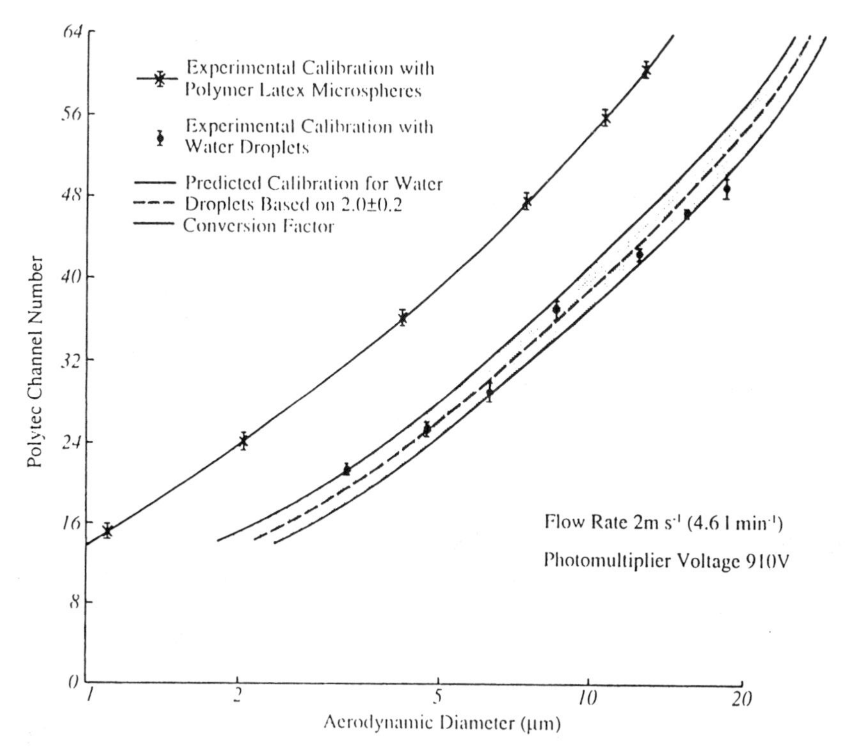

Figure 30 *Calibration of a Polytec HC-15 optical particle counter with PSL particles and water droplets*

number concentration obtained by counting the particles collected on selected areas of a membrane filter (Mitchell (1984) and Rimberg (1979)). It is important to obtain a homogeneous aerosol deposit on the membrane filter, and this can easily be achieved by mounting the main filter on top of a second. Although this procedure enables the efficiency of the OPC to be checked against an independent standard, it is time consuming unless an automated image analysis system can be used to scan the filter. Therefore, many workers prefer to obtain comparative performance data for an OPC against other similar instruments. The condensation nuclei counter can be used as an alternative to filter collection-microscopy (Kruger and Leuschner (1978)), especially if the OPC is being used to sample sub-micron particles.

5.3.2 Laser Diffraction Instruments

A range of instruments based on the principle of laser diffraction has been developed from the original work of Swithenbank *et al* (1977). While early instruments made use of the Fraunhofer approximation to resolve the scattered light intensity data to a particle size distribution, most analysers employ the full theory of angular light scattering developed by Mie.

An expanded laser beam produces a parallel beam of coherent, monochromatic light, and a Fourier transform lens is used to focus the diffraction pattern generated by the aerosol ensemble in the measurement zone onto a photodetector consisting of a series of concentric ring-diodes (Figure 31). Undiffracted light is reduced to a spot at the centre of the detector plane, with the diffraction pattern forming circular rings. The multi-element detector determines the radial distribution of diffracted light intensity, which can be related to the original size distribution by means of data inversion. Early instruments produced data fittings to continuous functions such as the Rosin–Rammler distribution, but other procedures have been developed which enable a more realistic 'model-independent' solution to be applied. These instruments are generally unsuitable for measurements of aerosols consisting of uniform, monodisperse particles.

Movement of the particles in the beam does not affect the diffraction pattern, since the light diffracted by a particle at a given angle will give the same radial displacement in the focal plane of the lens irrespective of the position of the particle in the beam. Thus, laser diffractometers are used to determine the size distribution of a population of particles weighted by volume, rather than by counting individual particles that enter a fixed measurement volume to provide a number–size distribution. They are not susceptible to coincidence problems, and can be used to measure concentrated aerosols and sprays. However, at very high concentrations, particle–particle scattering will lead to errors due to obscuration, and these instruments cannot be used reliably when this problem exceeds about 50%.

There is a wide range of commercially available laser diffractometers, that have mainly been developed for powder and spray analysers (Kaye (1979)). Not all of these instruments can be used directly with aerosols, and advice should be sought from the manufacturers. An outstanding feature with many of these instruments

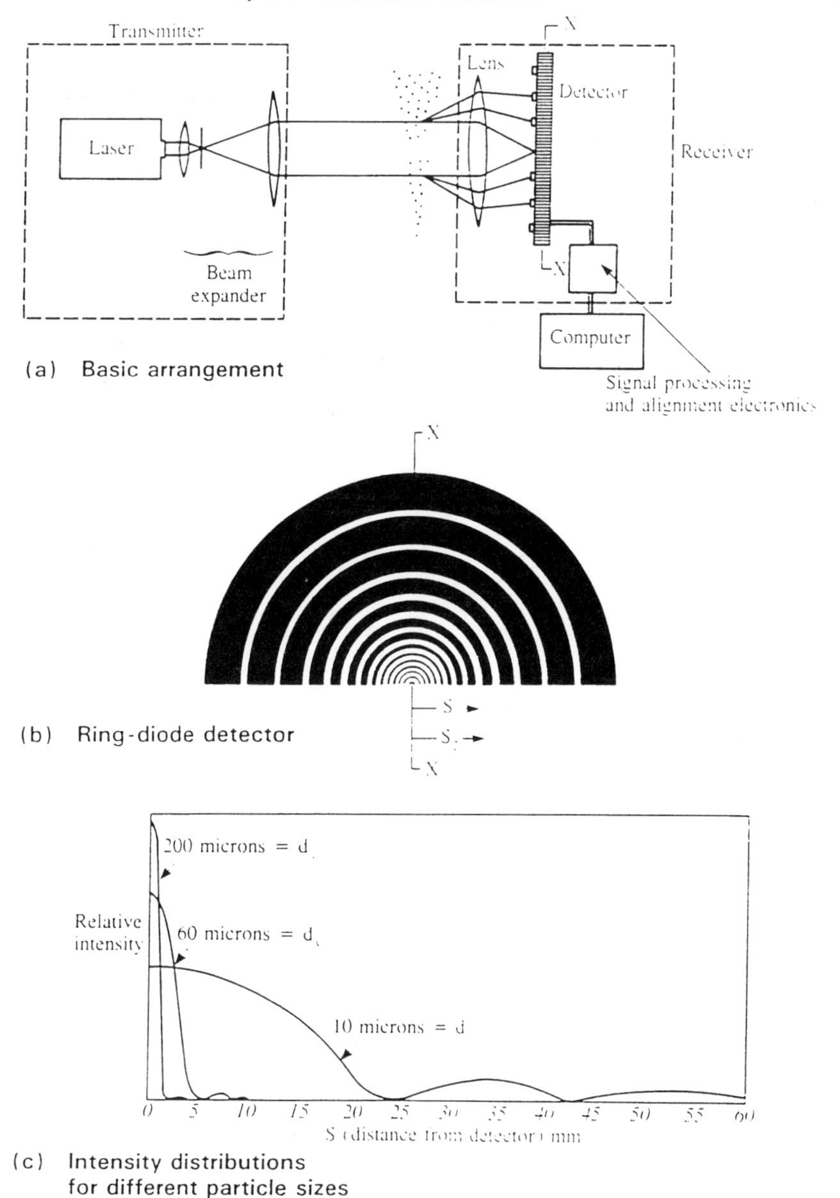

Figure 31 *Laser diffractometer—principle of operation*

is their ease-of-use, combined with their capability to make non-intrusive measurements (*e.g*. Malvern 2600). The optical components can be mounted in the open for ambient measurements or outside of the chamber containing the aerosol, and the processes observed by means of optically clear windows.

Laser diffractometers have a wide operating size range. Two or more orders of magnitude is typical, although additional lens combinations may be required to span the entire range of a given instrument. These systems do not consider individual aerosol particles, but produce a diffraction pattern-generated size distribution of a whole population. Calibration is carried out by the manufacturer, but validation/performance checks should periodically be made using a selection of monodisperse polystyrene latex particles.

5.3.3 Laser–Doppler Analysers

Other non-invasive particle sizing instruments are based on the laser–Doppler effect (Figure 32). The scattered light is detected as individual particles traverse a series of interference fringes formed from intersecting laser beams (Yeoman *et al* (1982)). This technique was developed from laser–Doppler anemometry, and

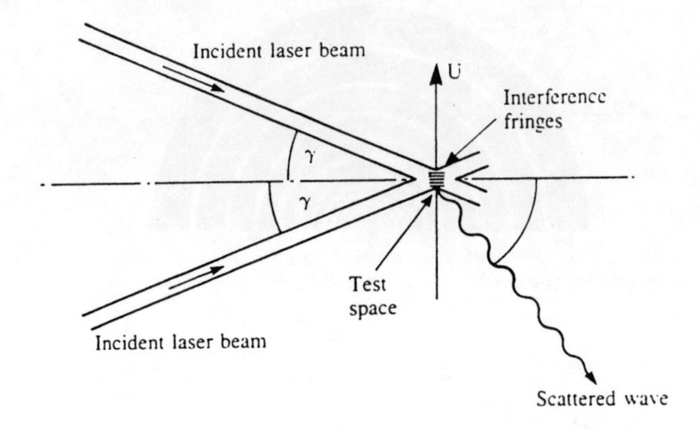

(a) Operating principle

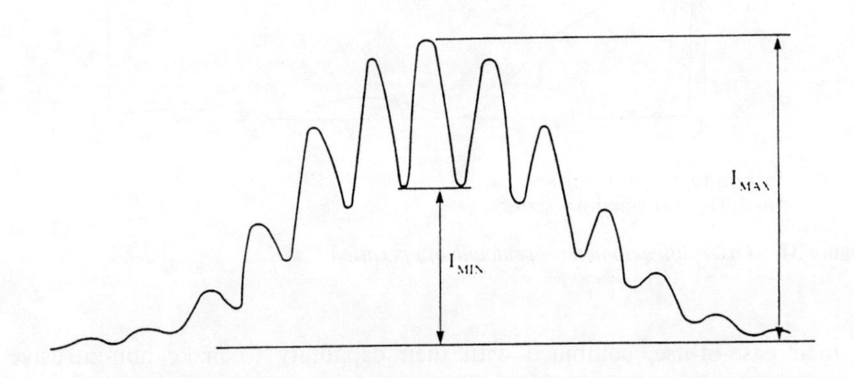

(b) Signal visibility profile

Figure 32 *Laser–Doppler particle sizing—operating principles*

individual particle velocity as well as size can be determined simultaneously. Aerosol particles in the size range 20 to 500 μm diameter and particle concentrations approaching 10^6 particles cm^{-3} can be sampled, although care must be taken to minimise coincidence effects. Checks on instrument performance accuracy are recommended as in Section 5.3.2, above.

5.3.4 Laser Intensity Deconvolution Analysers

Single-particle counting can be accomplished by measuring the absolute intensity of light scattered by particles traversing a focused laser beam. The Insitec PCSV family of instruments was developed from the work of Holve and Self (1979 and 1980) and Holve (1980) to measure the absolute intensity of light scattered as particles pass through an optically defined measurement volume (Figure 33). Since the sample volume has a non-uniform light intensity distribution, the intensity of the scattered light depends on both the particle size and trajectory. The potential ambiguity in particle size is resolved using a deconvolution algorithm based on two requirements:

(a) absolute scattered light intensity of a large number of particles must be measured (typically requiring less than 1 min),
(b) sample volume intensity distribution must be known.

The count-based spectrum of scattered light intensities is deconvoluted to yield the absolute particle number concentration as a function of size. Insitec instruments are fully non-invasive light scattering techniques, claimed to be capable of determining particle concentrations up to 10^7 particles cm^{-3} (well in excess of conventional single particle counters); the claimed size range of 0.2 to 200 μm is also very wide, but not all of this range may be directly available without making adjustments to the optical components. These analysers have been used to study gas-borne particles, slurries and powders, and measurements

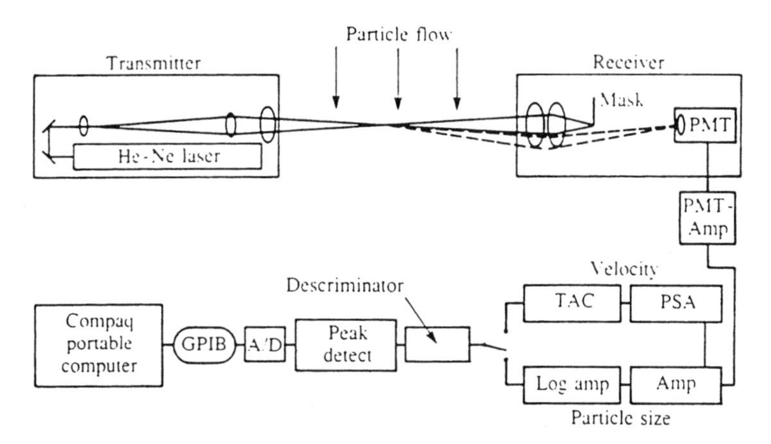

Figure 33 *Intensity deconvolution particle sizing technique*

have been made in arduous conditions, such as within fossil fuel combustion plant (Holve and Self (1980)). Furthermore, the performance of this instrument has been compared favourably with in-stack impactor measurements of emissions from a pilot-scale combustion plant (Dahlen *et al* (1987)).

5.3.5 Laser–Particle Interaction/Image Analyser

Laser–particle interaction/image analysers such as the GALAI-CIS analyser combine light blockage with image analysis in a single analyser (Karasikov *et al* (1988)), that can be used with almost any type of particle dispersion. The claimed operating size range for powders in liquid suspension is very wide (0.5 to 1200 μm), but not all of this is available in any one measurement, and the four ranges:

(a) 0.5 to 150 μm,
(b) 2 to 200 μm,
(c) 5 to 600 μm,
(d) 10 to 1200 μm,

are provided with the standard instrument.

The laser-interaction part of the analyser is based on the principle of time-of-transition, and has a dynamic range of 300:1. A focused low-power He-Ne laser beam of 1 μm width is scanned in a circle through the measurement zone which contains the particle suspension (Figure 34). This laser beam interacts with individual particles by light blockage, and a PIN photodiode detector is located directly behind the measurement area and senses the duration of each interaction. The width of the pulse represents the time of interaction of the laser beam with the particle as the laser beam scans across the surface, providing a parameter that is related to particle size and is independent of particle refractive index, the attenuation of the fluid supporting the particle, and the output power of the laser. The size resolution is excellent (300 channels per range selected), and is limited by the sampling rate of the detecting electronics.

Although the time–size relationship is unique, the laser beam diameter varies along the length of the beam, and ambiguity might arise from particles that interact with the laser beam out of the focal plane. However, such interactions produce longer pulse lengths for a given particle size, and the analyser can discriminate against these signals by pulse editing techniques. The laser beam may not interact with particles along their diameter or longest dimension, producing a signal that is shorter in duration than expected for a given particle size. This problem is easily resolved with spherical particles, since the rise and decay times of the pulse are steepest for an interaction crossing the centre of the particle and therefore the true diameter is measured; similar criteria can also apply with certain non-spherical, regular-shaped particles (*e.g.* spheroids). However, there is an inherent ambiguity in the interpretation of signals from irregularly shaped particles, and it is for this reason that the GALAI-CIS is equipped with an optical microscope coupled to a CCD TV camera-image

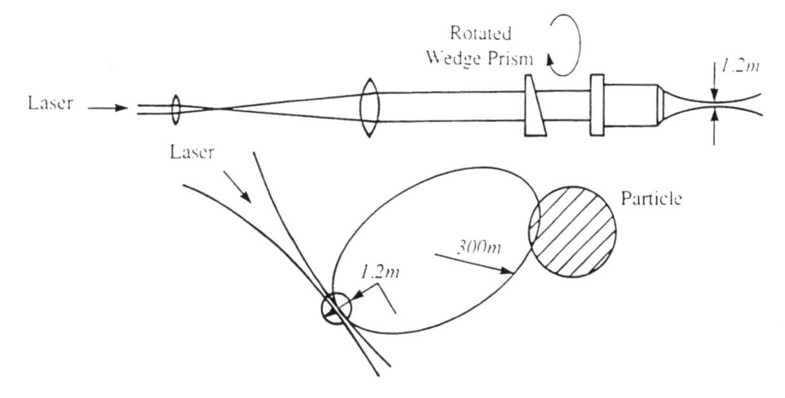

(a) Scanning laser beam for time-of-transition interaction with particles

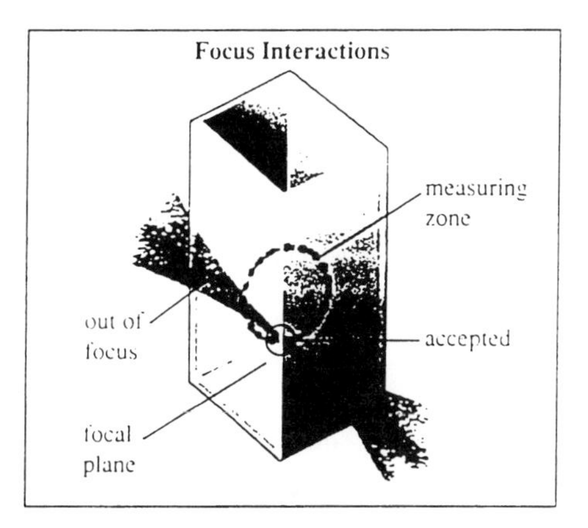

Out-of-focus or off-centre interactions generate rise times that are too long in proportion to the particle diameter. Using an algorithm based on the overall pulse signature with normalized rise-time criteria, each interaction is accepted or rejected. Only data from accepted signals are filed for analysis.

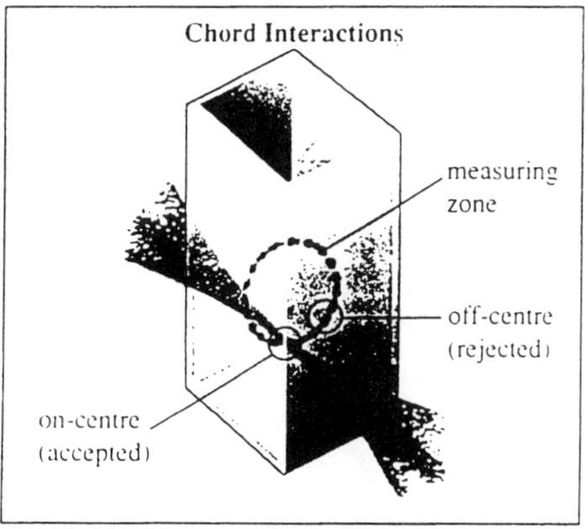

(b) Discrimination between focussed and off-centre particle-laser beam interactions

Figure 34 *GALAI-CIS-1 aerosol analyser*

analyser. The microscope optics are arranged orthogonally with respect to the axis of the laser interaction system. The manufacturer recommends that this facility be used to derive shape-related data for the particles (Aharonson

(1986)), and a powerful suite of image analysis software is available to interpret the various aspects of particle shape and size. However, it is not easy to interpret the size distribution data that originate from the laser interaction part of the equipment, particularly for complex-shaped particles.

The GALAI-CIS-1 analyser is designed to study both aerosols and liquid-borne particulates. Aerosols can be sampled directly using a purpose-built nebuliser/atomiser (Aharonson *et al* (1986) and Aharonson (1986)), and particles collected on membrane filters can also be examined by the image analysis mode (Karasikov and Krauss (1988)). Studies using the time-of-transition mode (Karasikov and Krauss (1989)) with the irregular-shaped certified reference materials (CRMs) dispersed in liquid from the European Bureau of Community Reference (Quartz CRMs 67, 68 and 69) indicate good agreement over a very wide size range (2 to 630 μm volume equivalent diameter) with the certification data (that are based on sedimentation). The time-of-transition analysis produces data that are not dependent upon particle refractive index; thus, coloured micron-sized monodisperse polystyrene latex microspheres can be sized correctly by both time-of-transition and image analysis modes. However, it should be noted that the measured monodispersity of these polymer latex particles is not as good as might have been expected from the specifications for these calibrants. As with some of the other laser-based techniques, calibration can be checked with a series of monodisperse polystyrene latex particles in liquid suspension or aerosol form.

5.4 Electrostatic Classifiers

Electrostatic classifiers distinguish between the electrical mobilities of singly charged particles in an applied electric field, and the most commonly used aerosol analysers of this type are derived from an instrument developed by Whitby and Clarke (1966). Three configurations are commercially available (all from Thermosystems Inc): Electrical Aerosol Analyser (EAA), Differential Mobility Particle Sizer (DMPS) and Scanning Mobility Particle Sizer (SMPS). The performance of the EAA was first described by Liu and Pui (1975), the DMPS by Knutson and Whitby (1975), and the SMPS by Wang and Flagan (1990). EAA comprises an electrostatic classifier with an integral electrometer, separate controller and an optional computer for control and data analysis (Figure 35). Both the DMPS and SMPS consist of an electrostatic classifier and separate condensation particle counter (Figure 36) together with electronics interface, interconnecting pipework and a computer for control and data interpretation (Keady *et al* (1983) and Flagan *et al* (1993)). The DMPS may also be used with an electrometer in place of the condensation particle counter.

All three systems can be calibrated with either monodisperse or polydisperse aerosols. The calibrant particles must be either solid or a low-volatile liquid, and their number concentration has to remain constant throughout the calibration period, which may be as long as 0.5 h when using dilute aerosols with the DMPS. Techniques for generating suitable test aerosols to calibrate the EAA have been described by Pui and Liu (1979), and may also be used with the

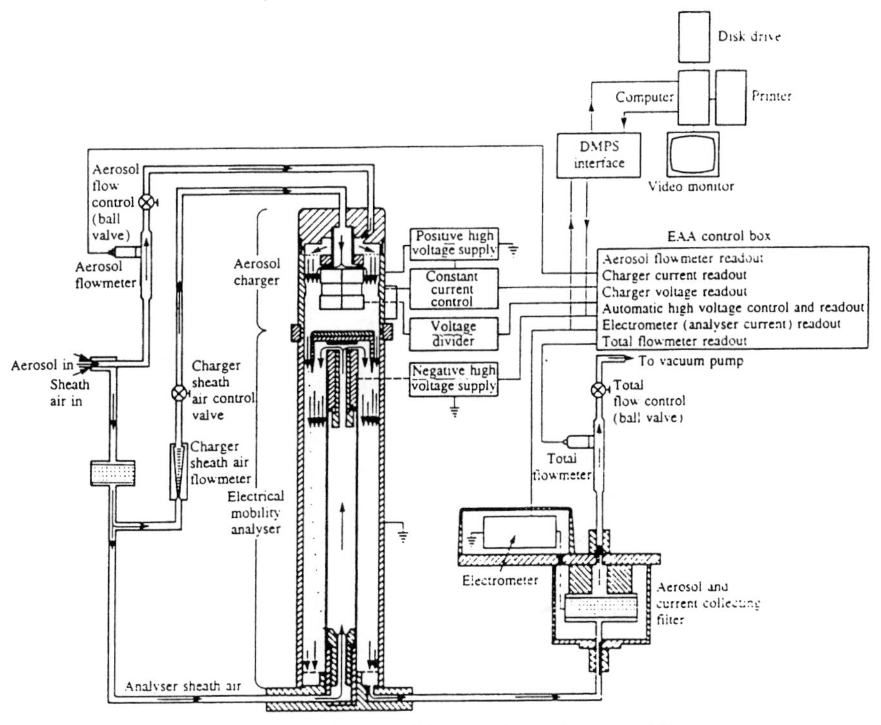

(a) Electrical Aerosol Analyser (TSI)

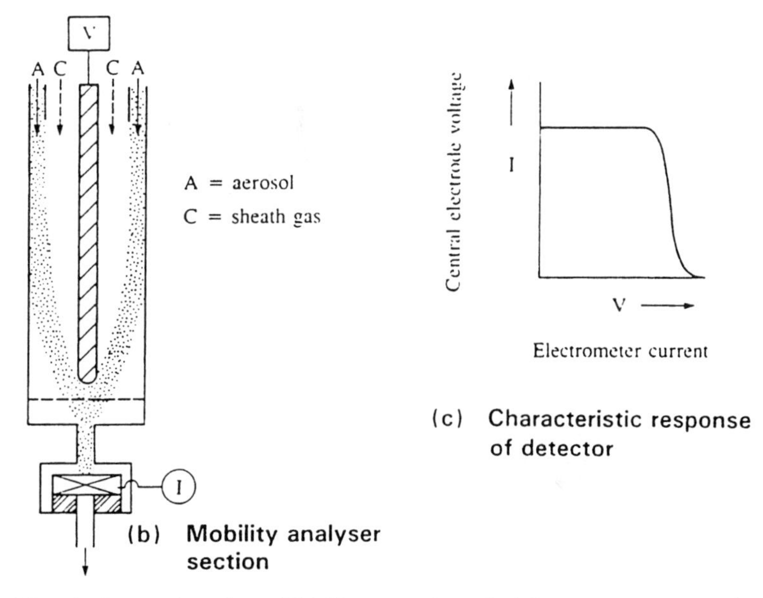

(b) Mobility analyser section

(c) Characteristic response of detector

Figure 35 *Electrical aerosol analyser (EAA)—operating principle*

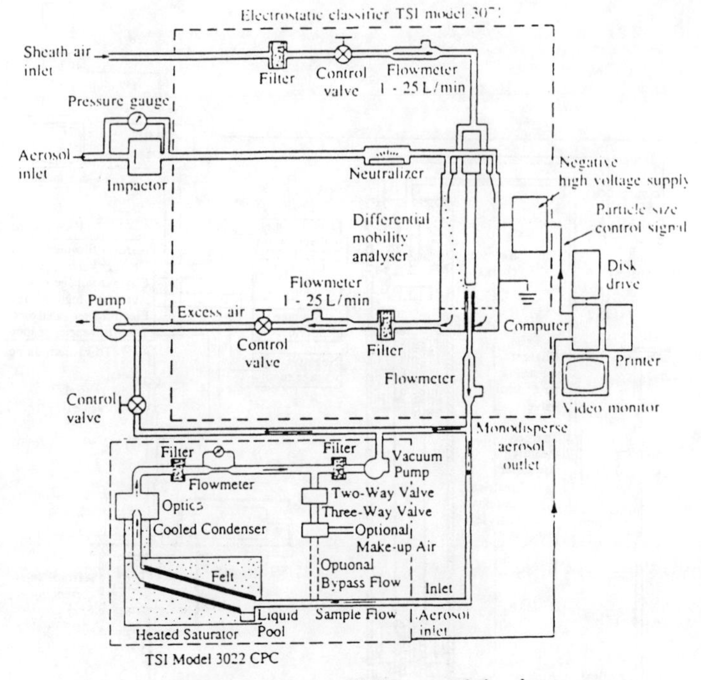

(a) Differential mobility particle sizer

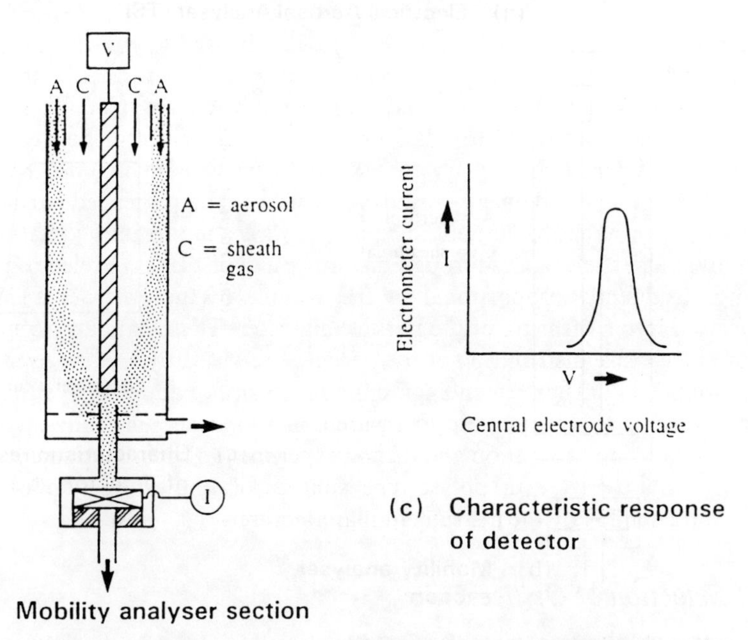

(b) Mobility analyser section

(c) Characteristic response of detector

Figure 36 *Differential mobility particle sizer (DMPS)—operating principle*

DMPS and SMPS. The electrical classification procedure can be used to produce monodisperse calibrants from polydisperse systems generated from nebulisers or gas-phase reactions. Dilute aqueous solutions of sodium chloride have also been atomised to produce aerosols that are suitable in demonstrating the equivalence of measurements made by the EAA and DMPS (Horton *et al* (1989)).

A major criticism of the calibration procedure is that the same electrical mobility technique is also used to produce the test aerosol. However, other methods of generating sub-micron calibrants cannot meet the stability and monodispersity requirements, especially when the CMD is smaller than 0.1 μm. At some stage in the calibration procedure, it is good practice to measure the size distribution of the test particles by an independent method, such as filter collection-microscopy.

Electrostatic classifiers are frequently calibrated against each other, and this approach has been chosen by the manufacturer for maintenance purposes (Sem (1979)). The response matrix of one electrical aerosol analyser is compared with a standard instrument sampling the same polydisperse aerosol (tobacco smoke or metal fume). This type of comparison is easy to perform and can be used to detect the anomalous behaviour of electrostatic classifiers on a routine basis.

5.5 Electrical Sensing-Zone Instruments

Although not strictly an aerosol analyser, electrical sensing-zone (ESZ) instruments such as the Coulter Counter are commonly used to size-analyse particles collected from an aerosol. They make use of electrical resistance to measure particle size distributions in liquid suspensions. A stream of the suspended particulate is constrained to pass through a small orifice of between 10 and 400 μm diameter, across which an electrical current is maintained by placing electrodes in the liquid on either side of the orifice. The suspended particles have a different resistance to that of the liquid, creating a momentary variation in the current as a particle passes through the orifice and displaces the electrolyte. This change is directly proportional to the volume of the particle, and can be converted into a measure of the particle diameter. These data are accumulated to produce a size distribution of the particles weighted by volume over the size range of 0.3 to 200 μm, depending on the size of orifice adopted. Wetting agents and ultrasonic energy are used to ensure that all particles are fully dispersed in the liquid during calibration and during experimental runs. Instrument calibration is carried out with polystyrene spheres of known size, following the procedure defined by the instrument manufacturer.

(a) Selection of Orifice Tube

An orifice tube must be selected with an aperture such that the expected maximum particle diameter is approximately 40% of the aperture diameter for most models of Coulter Counter, or up to 60% for the Model ZM and the Multisizer. More than one size of orifice tube may be required if a considerable amount of the sample is smaller than 2% of the aperture (the approximate size range of the sample can often be quickly determined by means of optical microscopy).

(b) Selection of Electrolyte

An electrolyte must be used that is compatible with the characteristics of the material to be analysed, and should be filtered before use to minimise the background count. Phosphate-buffered saline solution is compatible with human blood cells, and can be used for a wide range of materials that are insoluble in aqueous solutions. Advice on the selection of other electrolytes is available from the instrument manufacturer.

(c) Sample Dispersion

An ESZ instrument will only give a size analysis of the particulate material presented to the aperture. Hence, the particles constituting the material under test should be fully dispersed in the electrolyte before analysis begins. Dispersion can be achieved by gentle mixing with a suitable dispersant, applying ultrasonic vibrations, and high speed blending and mixing.

5.5.1 Operational Principles

The ESZ instrument determines the number and size of particles suspended in an electrically conductive liquid by forcing the suspension to flow through a small aperture that has an immersed electrode on either side. Each particle traversing the aperture changes the resistance between the electrodes to produce a voltage pulse of short duration which is proportional to particle size. The series of resulting pulses is electronically scaled and counted.

The voltage pulses are amplified and fed to a threshold circuit which has an adjustable threshold level. If this level is reached or exceeded, the pulse is counted. By taking a series of counts at selected threshold levels and correcting for coincidence, data are directly obtained to derive cumulative frequency plots versus particle size. Integration of all or part of the resulting curve provides a measure of the particle content of the suspension. The pulse height and instrument response are essentially proportional to the particle volume and fluid resistivity, and the measurements are expressed in terms of spherical particle diameters (*i.e.* unable to discern particle shape).

Although the combination of both liquid flow and electrical current through the aperture results in a complex signal for analysis, certain approximations can be made to simplify this procedure. Thus, the theoretical basis of the ESZ technique is relatively simple and 'side-effects' are notably few in number and small in effect. The accuracy of a measurement is not limited by the instrument, but is determined by the degree of care exercised in such studies. The instrument is inherently simple to calibrate, and requires only a laboratory balance and pipette.

5.5.2 Calibration of the Threshold Scale

Calibration of the ESZ instrument in absolute terms can be carried out by the count-integration procedure on narrow distributions of smooth particles of known density. Accurate checks can be quickly made with monodisperse particles that can be obtained from many sources. The following materials are available from Coulter Electronics for the calibration by number or weight (instructions for full-field calibrations are supplied by the manufacturers for each type of instrument):

> Latex suspensions are available in a range of nominal sizes and are designated for the calibration of specific sizes of aperture. With the exception of the 175 μm diameter spheres, these particles are suspended at a specified concentration in distilled water (containing surfactant and preservative) such that one to five drops are required in approximately 200 μl of electrolyte to calibrate an orifice tube approximately ten times the nominal latex diameter. Recommended concentrations are given with each vial, as are the various assayed diameters under different measurement conditions. Each vial of suspension is sufficient for at least 50 calibrations.
>
> Latex calibration suspensions are made from polystyrene latex, polyvinyltoluene latex, and polystyrene divinyl benzene (DVB) latex. The latter are recommended in particular, as they will not readily change size upon immersion in alcohols, ketones or any aqueous electrolyte. As with the measurement of real samples, calibration materials must be properly dispersed in the electrolyte.

Each vial of calibration material supplied by Coulter Electronics carries an assay sheet expressing median and modal diameters of the particles, determined in standard Coulter Isoton II Electrolyte solution. The procedure used to assay these sizes is described by Harfield and Wood (1971).

5.6 Optical and Electron Microscopy

Although microscopic techniques are labour intensive, they should not be neglected since they provide the only direct method of viewing particles with sufficient resolution to assess shape and surface structure as well as size. Microscopy represents a wide-ranging method of analysis, and an in-depth description of all of the available techniques is beyond the scope of this document. Emphasis is placed instead on the preparation of suitable samples of aerosol particles for microscopic inspection.

Microscopy remains the most valuable tool for particle size analysis because such measurements relate directly to the physical dimensions of the particles. A full description of optical microscopy for particle characterisation is given in part 1 of the Particle Atlas (McCrone and Delly (1973)), and basic measurement techniques are also described in Microscopy Handbook 23 of the Royal Microscopical Society (Bradbury (1991)). Suffice it to say, optical microscopes may be calibrated at known magnifications using the appropriate graticule. Basic guidelines for particle size analysis by optical microscopy can be found in BS 3406 : Part 4 (1993) and US ASTM Standard D-2009–65 (re-approved 1979). The British Standard contains detailed guidelines on the use of automated image analysers, which reduces some of the tedious operations associated with particle size analysis (Germani and Buseck (1991)).

Optical techniques are confined by the limit imposed on resolution due to the wavelength of the light source (*i.e.* particles larger than 3 μm projected area diameter). Smaller particles can be sized using electron microscopy, but such techniques operate in vacuum and may be inappropriate for volatile particles. At present, there is no satisfactory technique for handling sub-micron volatile particles, although cryogenic stages can be fitted to electron microscopes in an attempt to minimise evaporation. Definition of the exact magnification of an image produced in a scanning electron microscope (SEM) and a transmission electron microscope (TEM) is carried out by comparing the dimensions of the image with that of an appropriately lined carbon film replica of a diffraction grating under the same conditions as the particles.

If the particles are compatible with the conditions in an electron microscope, there are a growing number of electron-beam techniques available for probing the chemical composition on the surface and within the particle (Spurny (1986)). Combined physico-chemical analysis of particles is likely to become as important as particle sizing alone, since such a combination of information assists considerably in the understanding and control of chemical processes (Nichols and Bowsher (1988)).

Optical microscopy involves the recommended use of cellulose ester membrane filters (particles larger than 5 μm projected area diameter) since, after the particles have been collected, the opaque membrane can be cleared by exposure to the vapours of either acetone or a 50% *v*/*v* mixture of 1,4-dioxan and dichloromethane. However, these membranes (and glass fibre filters) are not a good choice for particle sizing by SEM, as their fibrous surface is sufficiently rough to merge with the outlines of the particulate debris. The merging of image outlines is especially problematic if automated image analysis methods are being used to size the particles, since it may not prove possible to distinguish the particle contours from the background, even when using instruments with 256 grey levels of discrimination. Over the previous 20 years, porous Nuclepore polycarbonate membranes (Nuclepore Corp.) have become widely accepted as the standard collection surface for use in the SEM, since they are microscopically smooth and result in clear images of the particles. It should be noted that these membranes are delicate and must be mounted on a support filter (usually cellulose acetate) to prevent the clustering of particles around the pores. More recently, a new filter based on porous alumina has become available (Anotec Separations): particles are not as easily distinguished, but the pressure drop across Anopore membranes is much lower, they are more efficient particle collectors (larger area filters and greater sampling flow rates), and can be used at higher temperatures. All of these filter media are non-conductive, and must be treated to prevent build-up of electrostatic charge in the SEM, with subsequent loss of image quality. The usual procedure is to apply a thin film of gold to the surface of the filter membrane under vacuum in a sputter-coating device prior to examination; a carbon coating is suitable for elemental analysis but is not recommended for particle sizing, since image quality is often poor.

Between 300 and 1000 particles may need to be sized to obtain representative data, depending on the required size resolution and the width of the size distribution. Such an exercise may require the inspection of as many as 100 images, which can be tedious unless automated (Allen (1981)). If an automated image analyser is used, care is required to ensure that the particles are always in focus and that the discrimination techniques used to outline the particles (*e.g.* outline erosion and space filling within outline) do not introduce systematic biases. Particles located on the boundary of the image should be eliminated from the analyses; most image analysers are equipped with software to perform this function automatically by introducing a narrow 'guard' area surrounding the main survey area. Detailed aspects of image analysis and particle shape and size have been reviewed in a book edited by Beddow (1984).

Recommended magnifications in the SEM are based on the expression:

$$N = 10^8 \left[\frac{l_m^2 A_f N_m}{m_m^2 L_m W_m n_m} \right]$$

where N is the total number of particles forming the size distribution, N_m is the number of particles counted per micrograph, m_m is the calibrated length of the micron-scale marker on the micrograph (μm), l_m is the actual length of this marker (cm), L_m and W_m are the length and width respectively of the micrograph (cm), n_m is the number of micrographs examined, and A_f is the collection area of the filter (cm^2).

5.7 Diffusion Batteries and Denuders

Diffusion batteries consist of a single stage or several stages in series. The diffusion cell in each stage can be a bundle of collimated round or rectangular tubes, or a stack of wire screens. Aerosol concentrations are measured sequentially between each stage and the aerosol size distribution is deduced from a knowledge of the size penetration characteristics of each stage. The calibration of a diffusion battery involves the determination of the penetration efficiency of particles of known size through each of the individual stages, which is usually achieved by measuring aerosol number concentrations before and after each stage using a condensation particle counter. Diffusion batteries are applicable to particles in the range 0.002 to 0.2 μm diffusional diameter, and multiple units can be operated in series or parallel to acquire particle size distribution data.

Unlike an impactor stage, a diffusion battery does not produce a sharp cut-off in size, but rather a gradual cut-off spanning more than one order of magnitude in particle diameter. A single measurement of a polydisperse aerosol with a diffusion battery yields an average diffusion coefficient that can be converted to other types of average diameter. Several diffusion batteries with different effective lengths or the same one at different flow rates can be used to obtain a series of penetration measurements, and computer-based inversion schemes are available that will give the best-fitting size distribution to a series of diffusion battery data (Knutson and Sinclair (1979)).

Monodisperse calibration aerosols for diffusion batteries are typically generated by the nebulisation of preformed polymer latex particles and/or vapour condensation, followed by classification using a calibrated electrical aerosol analyser. Large fluctuations in aerosol concentration must be avoided by generating the calibration aerosol into a relatively large enclosure before sampling with the battery.

Users must ensure that they operate a diffusion battery at the manufacturer's specified flow rate, which should be determined using a traceable technique. Inlet gas temperature, relative humidity and laboratory pressure must be measured to ensure that the battery is operated at the same standard flow rate as during calibration. However, it is difficult to demonstrate the traceability of the calibration of a diffusion battery to international standards due to the complexity of the data inversion techniques used to calculate the aerosol size distribution and the inability to demonstrate that the condensation particle counter detects the calibrant particles with an efficiency of 100%.

5.8 Condensation Particle Counters (CPCs)

While condensation particle counters cannot measure particle size distributions directly, they are frequently used in conjunction with other instruments to achieve this objective (*e.g.* electrostatic classifiers and diffusion batteries). Primarily, CPCs have been developed to measure the number concentrations of sub-micron particles, and include manually operated systems such as the Nolan-Pollack CPC (Nolan and Pollack (1946)) and the continuous flow CPC (Agarwal and Sem (1980)). The Nolan-Pollack CPC is often used as a secondary standard for the calibration of other CPCs (Pollack and Metnieks (1959)) because the decrease in light transmission across the expansion chamber is directly related to the number of particles.

CPC calibrations rely on the production of nuclei of known number concentration, that can be counted either by direct microscopic observation, or from photographs of the droplets formed in the CPC. Alternatively, the droplets can be deposited on a coated glass slide and be counted using a microscope. The continuous flow CPC has been developed by Agarwal and Sem (1980) and can be calibrated by counting the number of particles of known size and number concentration produced by an electrostatic classifier (Liu and Pui (1974)). Number concentration measurements are made using a Faraday-cup electrometer, in which multiply charged particles with the same electrical mobility as the calibration aerosol constitute a source of error that can be eliminated by applying a correction factor.

The continuous-flow CPC can be used over a wide range of number concentration (from 1 to 10^7 particles cm^{-3}) and operates in two different modes: individual particles are counted when the concentration is below 10^3 particles cm^{-3}, while above this value the total light extinction caused by the droplets in the 'view volume' (photometric mode) is measured. A correction factor is applied in the count mode to compensate for coincidences as concentrations approach the upper limit. When calibrating the system in the count mode, monodisperse aerosol from the electrostatic classifier is diluted with clean excess air to reduce the particle concentration to the desired range. The amount of dilution air is gradually reduced until only the monodisperse aerosol stream is sampled, and calibration data have been obtained for concentrations in excess of 5×10^5 particles cm^{-3} (Agarwal and Sem (1980)).

It is important to replace the n-butanol working fluid in the continuous-flow CPC at regular intervals: water vapour is absorbed from the air during use, diluting the n-butanol and reducing the size of the droplets formed in the condensation chamber. This source of error arises when the CPC is operated in the photometric mode, and is most significant at the highest aerosol concentrations.

CHAPTER 6

Concluding Remarks

6.1 The choice of a particular calibration method depends largely on the type of particle size analyser being calibrated and the degree of accuracy required. Emphasis has been placed in this manual on the appropriate calibration procedures for equipment that measure the aerodynamic diameter and the corresponding particle size distribution, although some consideration has also been given to such functional parameters as electrical mobility, light-scattering and diffusion coefficients that are used to quantify particle size.

Care should be taken to achieve the optimum conditions for the generation and sampling of the calibrant aerosol(s). Calibrations using monodisperse particles are more frequently carried out than those with polydisperse aerosols despite the extra work involved, since the uncertainties introduced by the monodisperse technique are judged to be smaller. This is particularly true with impactors and cyclones where the ECD values are determined from collection efficiency curves that are usually sensitive functions of particle size. High resolution aerosol size analysers such as the APS3320, API, most OPCs and electrical aerosol analysers are also best calibrated with monodisperse particles to determine whether changes have occurred in their calibration functions, which may indicate the onset of a degradation in performance. However, calibrations using polydisperse aerosols have advantages when applied to inertial spectrometers (which collect the entire aerosol for subsequent examination).

Other physical properties of the calibrant are often as significant as the degree of monodispersity. For instance, the refractive index is important when calibrating an OPC, and the choice of solid or liquid particles may influence the validity of APS3320 and API calibrations. The inertial pre-collector method is advantageous for OPCs, thus avoiding the need to know the refractive indices of either the calibration aerosol or the particles under study. When an APS3320 or API is being used to monitor liquid droplets, it may be necessary to refine the calibration technique so that the performance can be measured with a liquid that has similar surface tension and viscosity. Particle density, porosity and shape also have a marked influence upon the sensitivity of aerodynamic particle size analysers.

6.2 It is recommended that every aerosol analyser is calibrated on a regular basis as part of a QA-checking and maintenance policy.

6.3 The performance of ancillary equipment such as gas flowmeters, pressure gauges, *etc.* is important, and these systems need to be regularly calibrated prior to use.

6.4 This manual is viewed as a developing document which will continue to evolve with the advent of feedback from users. A primary aim is to create an easy-to-use handbook that covers a wide range of aerosol instrumentation for a variety of measurements. Generic information is also provided in the text, with the provision of background references for further study. The document represents the provision of a practically orientated manual that focusses on particle sizing, and ranges from the most rigorous tests (with full traceability) to day-to-day functional studies.

Appendix A: Cocktail Reference Materials for Aerosol Analysis

A.1 Introduction

A series of four three-component cocktail reference materials containing equal number concentrations of each component have been procured as part of the VAM programme (Marshall (1995a)). The materials consist of spherical polymer latex particles with volume equivalent diameters in the range from 0.1 to 5.0 μm; their specifications, traceability and availability are described below.

A.2 Specifications

Each of the cocktails comprises the components listed in Table A.1 at number concentrations of 4×10^8 particles cm^{-3}. This number concentration was chosen to produce a singlet to multiplet ratio of 0.99 when the suspensions are converted to aerosol form by means of conventional medical nebulisers that produce sprays with droplet volume equivalent diameter and geometric standard deviation of 3 μm and 2.0, respectively.

A.3 Traceability

JSR, Japan produced the cocktail reference materials by blending appropriate quantities of their existing range of monodisperse polymer latex products. Particle diameter traceability was achieved using the measurement technique described by Katsuta *et al* (1987), which involves the use of a transmission electron microscope calibrated with a catalase crystal sized absolutely by X-ray diffraction analysis. This method addresses all sources of systematic and random uncertainty (*i.e.* discrepancy between actual and nominal microscope magnification, variations in calibrant diffraction grating, growth of particles due to operating conditions of the microscope, and particle shrinkage under the influence of the electron beam) to give a total measurement uncertainty of

Table A.1 *Cocktail components—volume equivalent diameters*

Cocktail	*Component volume equivalent diameters* μm
A	0.1, 0.2, 0.5
B	0.2, 0.5, 1.0
C	0.5, 1.0, 2.0
D	1.0, 2.0, 5.0

$\pm 1.5\%$. The results are in excellent agreement with measurements from an independent study by Kousaka *et al* (1988), and are considered to be extremely reliable for the determination of particle diameter. JSR guaranteed a three-year shelf life for the diameter of the particles, provided that the sample bottles remain unopened and are stored in a refrigerator or at room temperature; this guarantee does not extend to opened bottles. They also agreed to continue to monitor the sizes of the particles at regular intervals.

A.4 Availability

The cocktail reference materials are available as samples of aqueous suspension containing approximately 10^6 particles. Samples are marketed through the Office of Reference Materials of the Laboratory of the Government Chemist, and appeared in their catalogue of reference materials during 1996.

The cocktails are of immediate application to instruments that measure particle diameters and concentrations in the liquid phase. However, many users will also require the particles in the aerosol form, and pneumatic nebulisation will be the most commonly used generating technique. The cocktails as provided contain measured particle diameters that are equally applicable in the liquid-borne or aerosol states. However, the measured number concentrations cannot be easily transferred from the suspension to the aerosol form; the relative number concentrations of each cocktail component in the aerosol will depend on the concentrations in the initial liquid suspension and the type of aerosol generation technique. Booker and Horton (1995) have described a nebuliser-based aerosol generator which overcomes the majority of practical difficulties, and this device has been successfully used to nebulise samples of these cocktails.

Appendix B: Particle Shape Standards for Aerosol Analysis

B.1 Introduction

The need for particle shape standards has been assessed in terms of the requirements of the particle measurement community (Lewis *et al* (1993)). As a consequence of these user-defined needs, a series of three fibre-analogue standards were prepared as part of the VAM programme (Marshall (1995b)). The particles were manufactured by silica and silicon micromachining, which offers great promise as a means of forming fibre-analogue particles (Hoover *et al* (1990)). Particle shapes were precisely engineered using a technology adopted to prepare standard semiconductor wafers. The technique only produces milligram quantities of product and is comparatively expensive (£250 per vial containing approximately 10^6 particles). Although production batches are limited to micrograms, such preparations contain several hundred-thousand micron-sized particles, which is more than adequate to calibrate single-particle counting aerosol analysers. It is also believed that there is scope for unit cost reductions if the production output can be scaled up.

B.2 Specifications

Three fibre-based shape standards of constant three-dimensional shape were prepared at the UK centre of excellence for this technology (Engineering Research & Development Centre, University of Hertfordshire). The shape standards have an aspect ratio (length/width) in the range from 1:1 to 10:1, and have a constant width/depth ratio close to unity (width and depth close to 1 μm). These standards are of immediate use to the increasing number of users with single-particle counting instruments. The final specification was:

(a) fixed width and depth of 1.7 and 1.0 μm respectively,
(b) variable length from 3 to 12 μm in three steps (3, 7.5 and 12 μm).

Table B.1 *SEM-measured dimensions of particles*

	Length μm		*Width μm*		*Depth μm*	
Nominal length μm	*Mean*	*Standard deviation*	*Mean*	*Standard deviation*	*Mean*	*Standard deviation*
3.0	3.09	0.10	1.67	0.08	0.96	0.09
7.5	7.51	0.22	1.72	0.11	1.02	0.06
12.0	12.13	0.22	1.70	0.04	1.00	0.07

Table B.2 *Indicative aerodynamic diameters of fibre-analogue particles (estimated 5% uncertainty in the indicative values)*

	Indicative aerodynamic diameter μm	
Nominal particle length μm	*Motion perpendicular to major axis*	*Motion parallel to major axis*
3.0	2.89	3.14
7.5	3.54	4.11
12.0	3.78	4.51

The lengths and width were not chosen arbitrarily, but corresponded to photolithographic masks that had already been manufactured at the University of Hertfordshire. Thus, new masks were not required, which represented a significant reduction in production costs and increased the number of particles that could be obtained with the available resources. The depth is technique-dependent and therefore could not be specified exactly.

All of the particles are essentially rectangular with a constant cross section and rounded corners due to the etching process. The dimensions of the particles were determined under a calibrated scanning electron microscope, and are listed in Table B.1 in terms of their mean values and standard deviations. Coefficients of variation ranged from 1.8 to 3.2%, 2.4 to 6.4%, and 5.9 to 9.4% for the lengths, widths and depths, respectively. Calculated aerodynamic diameters were derived using the theory of Oseen (1927) for the motion of equivalent spheroids perpendicular and parallel to the major axis, as given in Table B.2.

B.3 Availability

The particle shape standards are available in samples of aqueous suspension containing approximately 10^6 particles. Samples are marketed through the Office of Reference Materials of the Laboratory of the Government Chemist, and appeared in their catalogue of reference materials during 1996. The most popular uses for these shape standards are expected to be as model particles for research purposes and as instrument calibration aids. There is a requirement to deliver the shape standards in a form suitable for sampling by aerosol analysers in both these applications, and the best results are achieved using a direct method to introduce the particles to the inlet of the analyser in question. Several drops of the aqueous suspension containing the particles should be placed on a clean glass microscope slide. The drops of suspension can be dispersed over the surface of the slide using propan-2-ol (isopropyl alcohol), and should be dried in a gentle flow of particle-free air. The slide containing the dry particle deposit should be placed at the inlet of the aerosol analyser under investigation, and the deposit resuspended using a fine brush. This technique will deliver particles to the immediate vicinity of the aerosol analyser inlet in a sufficiently high airborne concentration to ensure detection of a large proportion of the particles by the analyser. Calibrant samples for microscopic analysis may require further 'purification' by the addition of high-priority propan-2-ol, centrifugation and removal of the supernatant liquor, to be repeated a number of times before use.

The technique described above may not be appropriate in every application and the need may arise to generate the shape standard particles as an aerosol. A pneumatic medical nebuliser can be used to produce an aerosol from an aqueous suspension of the particle shape standard (neat or diluted), but the airborne concentration can be rather low and care must be taken to choose an appropriate containment. A volatile solvent (*e.g.* propan-2-ol) is recommended to dilute the neat suspension. This has the advantage that the solvent is easily removed from the particles once they are airborne, and eliminates the need to introduce any form of drying or conditioning which would further reduce an already low airborne particle concentration. There is also a possibility, particularly with the longer particles, that breakage could occur during this type of aerosol generation; the risk of such damage can be minimised by using the lowest possible flow rate through the nebuliser.

The manufacturer of the shape standard particles recommends that the suspensions should be maintained at neutral pH. These standards are also supplied in an aqueous suspension which does not contain any surfactant. Consequently, the suspended particles have a tendency to agglomerate. Particles will also sediment from the suspended sample, and collect at the bottom of the container with further opportunity for agglomeration. It is essential that the suspension is placed in an ultrasonic bath for up to 2 min prior to use. While there is no evidence that ultrasonication can induce particle fracture, mechanical stirring of the suspension should be avoided to prevent or minimise particle breakage.

References

J. K. Agarwal and G. J. Sem (1980), 'A Continuous-Flow, Single-Particle-Counting Condensation Nucleus Counter', *J. Aerosol Sci.*, **11(4)**, 343.

J. K. Agarwal, R. J. Remiarz, F. R. Quant and G. J. Sem (1982), 'Real-Time Aerodynamic Particle Size Analyser', *J. Aerosol Sci.*, **13(3)**, 222.

E. F. Aharonson (1986), 'Shape Analysis of Airborne Aerosol Particles by the GALAI-CIS-1', p. 398 in Proc. Ind. Inst. Aerosol Conf., Berlin, Germany, Eds.: W. O. Schikarski, H. J. Fissan and S. K. Friedlander, Pergamon Press, Oxford, UK.

E. F. Aharonson, N. Karasikov, M. Roitberg and J. Shamir (1986), 'GALAI-CIS-1: A Novel Approach to Aerosol Particle Size Analysis', *J. Aerosol Sci.*, **17(3)**, 530.

T. Allen (1981), 'Particle Size Measurement', 3rd Edn, Chapman and Hall, London, UK.

R. S. Babington, A. A. Yetman and W. R. Slivka (1969), 'Method of Atomizing Liquids in a Mono-dispersed Spray', US Patent 3 421 692.

M. R. Bailey and J. C. Strong (1980), 'Production of Monodisperse, Radioactively Labelled Aluminosilicate Glass Particles Using a Spinning Top', *J. Aerosol Sci.*, **11(5/6)**, 557.

P. A. Baron (1984), 'Aerodynamic Particle Sizer Calibration and Use', p. 215 in 'Aerosols', Eds.: B. Y. H. Liu, D. Y. H. Pui and H. J. Fissan, Elsevier Science, New York, USA.

P. A. Baron (1986), 'Calibration and Use of the Aerodynamic Particle Sizer (APS 3300)', *Aerosol Sci. Technol.*, **5(1)**, 55.

E. A. Barringer and H. K. Bowen (1982), 'Formation, Packing and Sintering of Monodisperse TiO_2 Powders', *J. Am. Ceram. Soc.*, **65(12)**, C-199.

J. K. Beddow (1984), 'Particle Characterisation in Technology: Volume II — Morphological Analysis', CRC Press, Boca Raton, Florida, USA.

R. N. Berglund and B. Y. H. Liu (1973), 'Generation of Monodisperse Aerosol Standards', *Environ. Sci. Technol.*, **7(2)**, 147.

D. B. Blackford and K. L. Rubow (1986), 'A Small-Scale Powder Disperser', p. 13 in Proc. NOSA Aerosol Symposium, Solna, Sweden.

D. B. Blackford, A. E. Hanson, D. Y. H. Pui, P. Kinney and G. P. Ananth (1988), 'Details of Recent Work Towards Improving the Performance of the TSI Aerodynamic Particle Sizer' p. 311 in Proc. 2nd Aerosol Society Conf., Bournemouth, UK.

D. R. Booker and K. D. Horton (1995), 'VAM14 — Annex B: (II), Development of Aerosol Concentration Standards', AEA Technology Report AEA-TPD-0345.

D. Boulaud, M. Diouri and G. Madelaine (1981), 'Parameters Influencing the Collection Efficiency of Solid Aerosols in Cascade Impactors', p. 125 in Proc. 9th Conf. European Association for Aerosol Research, Duisburg, Germany.

D. Boulaud, J. C. Chouard, C. Frambourt and G. Madelaine (1983), 'Study of a Sedimentation Battery', *J. Aerosol Sci.*, **14(3)**, 421.

S. Bradbury (1991), 'Basic Measurement Techniques for Light Microscopy', Microscopy Handbook 23, Royal Microscopical Society, Oxford University Press, UK.

H. Buettner (1985), 'Measurement of Particle Size Distributions in Gas Flows with an Optical Particle Counter', *Part. Charact.*, **2(1)**, 20.

R. D. Cadle and P. L. Magill (1951), 'Preparation of Solid- and Liquid-in-Air Suspensions for Use in Air Pollution Studies', *Ind. Eng. Chem.*, **43(6)**, 1331.

H.-C. Chang (1974), 'A Parallel Multicyclone Size-Selective Particulate Sampling Train', *Am. Ind. Hyg. Assoc. J.*, **35**, 538.

P. K. P. Cheah and C. N. Davies (1984), 'The Spinning-Top Aerosol Generator — Improving the Performance', *J. Aerosol Sci.*, **15(6)**, 741.

B. T. Chen and D. J. Crow (1986), 'Use of an Aerodynamic Particle Sizer as a Real-Time Monitor in Generation of Ideal Solid Aerosols', *J. Aerosol Sci.*, **17(6)**, 963.

B. T. Chen, Y. S. Cheng and H. C. Yeh (1985a), 'Performance of a TSI Aerodynamic Particle Sizer', *Aerosol Sci. Technol.*, **4**, 89.

B. T. Chen, H. C. Yeh and Y. S. Cheng (1985b), 'A Novel Virtual Impactor: Calibration and Use', *J. Aerosol Sci.*, **16(4)**, 343.

D. K. Craig, A. P. Wehner and W. G. Morrow (1972), 'The Generation and Characterisation of a Respirable Aerosol of Chrysotile Asbestos for Chronic Inhalation Studies', *Am. Ind. Hyg. Assoc. J.*, **33**, 283.

R. S. Dahlen, R. Beittel, B. A. Morgan and D. J. Holve (1987), 'Applications of an In-Situ Optical Counter for Combustion Measurements: Comparisons with Cascade Impactor Measurements', Proc. Meeting of the American Association for Aerosol Research (AAAR), Seattle, USA.

C. N. Davies and P. K. P. Cheah (1984), 'Spinning Generators of Homogeneous Aerosols', *J. Aerosol Sci.*, **15(6)**, 719.

R. Dennis (1976), 'Handbook on Aerosols', Technical Information Center, US Energy Research and Development Administration, USA.

R. T. Drew and S. Laskin (1971), 'A New Dust-Generating System for Inhalation Studies', *Am. Ind. Hyg. Assoc. J.*, **32(5)**, 327.

C. I. Fairchild and L. D. Wheat (1984), 'Calibration and Evaluation of a Real-Time Cascade Impactor', *Am. Ind. Hyg. Assoc. J.*, **45(4)**, 205.

H. J. Fissan and C. Helsper (1982), 'Techniques for Calibration of Dust Measurement Methods', VDI Berichte No. 429, Germany.

H. J. Fissan and C. Helsper (1983), 'Calibration of the Polytec HC-15 and HC-70 Optical Particle Counters', Chapter 58 in 'Aerosols in the Mining and Industrial Work Environments', Eds.: V. A. Marple and B. Y. H. Liu, Ann Arbor Science, Ann Arbor, Michigan, USA.

R. C. Flagan, F. R. Quant, K. D. Horton, L. M. Russell, G. J. Sem and M. M. Havlicek (1993), 'Scanning Mobility Particle Sizer: Characterisation and Intercomparison' p. 292 in Proc. 12th Annual Meeting of the American Association for Aerosol Research, Oak Brook, Illinois, USA.

J. P. Flesch, C. H. Norris and A. E. Nugent (1967), 'Calibrating Particulate Air Samplers with Monodisperse Aerosols: Application to the Andersen Cascade Impactor', *Am. Ind. Hyg. Assoc. J.*, **28**, 507.

H. Franzen and H. J. Fissan (1979), 'Collection Behaviour of Andersen Non-Viable Samplers and Andersen Stack Samplers Using Glass Fibre Collection Plates', *Staub-Reinhalt. der Luft*, **39**, 50.

H. Franzen, H. J. Fissan and U. Urban (1978), 'Calibration of an Andersen Stack Sampler (Mark III), Using the Berglund-Liu Aerosol Generator', *Staub-Reinhalt. der Luft*, **38**, 436.

N. A. Fuchs (1963), 'On the Stationary Charge Distribution on Aerosol Particles in a Bipolar Ionic Atmosphere', *Geofis. Pura. Appl.*, **56**, 185.

N. A. Fuchs (1964), 'Mechanics of Aerosols', Pergamon Press, Oxford, UK.

N. A. Fuchs (1973), 'Latex Aerosols—Caution!', *J. Aerosol Sci.*, **4**, 405.

N. A. Fuchs (1978), 'Aerosol Impactors (A Review)', Chapter 1 in 'Fundamentals of Aerosol Science', Ed.: D. T. Shaw, John Wiley and Sons, New York, USA.

N. A. Fuchs and F. I. Murashkevich (1970), 'Laboratory Powder Disperser (Dust Generator)', *Staub-Reinhalt. der Luft*, **30(11)**, 1.

N. A. Fuchs and A. G. Sutugin (1963), 'Droplet Size Distribution of Butyl Phthalate Mists Prepared by Vapor Condensation on Nuclei', *Kolloid Zh.*, **25(4)**, 487.

N. A. Fuchs and A. G. Sutugin (1966), 'Generation and Use of Monodisperse Aerosols', Chapter 1 in 'Aerosol Science', Ed.: C. N. Davies, Academic Press, New York, USA.

M. J. Fulwyler, J. D. Perrings and L. S. Cram (1973), 'Product of Uniform Microspheres', *Rev. Sci. Instrum.*, **44(2)**, 204.

J. A. Garland, A. C. Wells and E. J. Higham (1982), 'The Deposition of Particles in the Coolant Circuit of the Windscale Advanced Gas-Cooled Reactor', UK Atomic Energy Authority Report AERE-R 10491.

D. M. Garvey and R. G. Pinnick (1983), 'Response Characteristics of the Particle Measuring Systems Active Scattering Aerosol Spectrometer Probe (ASASP-X)', *Aerosol Sci. Technol.*, **2(4)**, 477.

M. S. Germani and P. R. Buseck (1991), 'Automated Scanning Electron Microscopy for Atmospheric Particle Analysis', *Anal. Chem.*, **63**, 2232.

A. Goetz, H. J. R. Stevenson and O. Preining (1960), 'The Design and Performance of the Aerosol Spectrometer', *J. Air Pollut. Control Assoc.*, **10(5)**, 378.

W. D. Griffiths, S. Patrick and A. P. Rood (1984), 'An Aerodynamic Particle Size Analyser Tested with Spheres, Compact Particles and Fibres Having a Common Settling Rate under Gravity', *J. Aerosol Sci.*, **15(4)**, 491.

W. D. Griffiths, P. J. Iles and N. P. Vaughan (1986a), 'Calibration of the APS33 Aerodynamic Particle Sizer and Its Usage', *TSI J. Particle Instrum.*, **1(1)**, 3.

W. D. Griffiths, P. J. Iles and N. P. Vaughan (1986b), 'The Behaviour of Liquid Droplet Aerosols in an APS 3300', *J. Aerosol Sci.*, **17(6)**, 921.

J. C. Guichard (1976), 'Aerosol Generation Using Fluidized Beds', p. 173 in 'Fine Particles', Ed.: B. Y. H. Liu, Academic Press, New York, USA.

J. S. Haggerty and R. W. Cannon (1981), 'Sinterable Powders from Laser-Driven Reactions', p. 165 in 'Laser Induced Chemical Processes', Ed.: J. I. Steinfeld, Plenum Publ. Co., New York, USA.

J. G. Harfield and W. M. Wood (1971), 'Standard Calibration Materials for the Coulter Counter', p. 293 in 'Particle Size Analysis 1970', Bradford, Society for Analytical Chemistry, London, UK.

C. O. Hendricks and S. Babil (1972), 'Generation of Uniform 0.5–10 μm Solid Particles', *J. Phys. E.: Sci. Instrum.*, **5**, 905.

W. C. Hinds (1982), 'Aerosol Technology: Properties, Behavior and Measurement of Airborne Particles', John Wiley and Sons, New York, USA.

D. J. Holve and S. A. Self (1979), 'An Optical Particle Sizing Counter for In-Situ Measurements: Parts I and II', *J. Appl. Optics*, **18(10)**, 1632.

D. J. Holve and S. A. Self (1980), 'Optical Measurements of Mean Particle Size in Coal-Fired MHD Flows', *Combustion Flame*, **37**, 211.

D. J. Holve (1980), 'In-Situ Optical Particle Sizing Technique', *J. Energy*, **4(4)**, 176.

M. D. Hoover, S. A. Casalnuovo, P. J. Lipowicz, C. Y. Hsu, R. W. Hanson and A. J. Hurd (1990), 'A Method for Producing Non-Spherical Monodisperse Particles Using Integrated Circuit Fabrication Techniques', *J. Aerosol Sci.*, **21(4)**, 569.

K. D. Horton and J. P. Mitchell (1990), 'The Calibration of Andersen Mark-II and California Measurements PC-2 Cascade Impactors', UK Atomic Energy Authority Report AEA-TRS-5001.

K. D. Horton and J. P. Mitchell (1992), 'The Calibration of a California Measurements PC-2 Quartz Cascade Impactor (QCM)', *J. Aerosol Sci.*, **23(5)**, 505.

K. D. Horton, R. D. Miller and J. P. Mitchell (1991), 'Characterisation of a Condensation-Type Monodisperse Aerosol Generator (MAGE)', *J. Aerosol Sci.*, **22(3)**, 347.

K. D. Horton, J. P. Mitchell and A. L. Nichols (1989), 'Experimental Comparison of Electrical Mobility Aerosol Analyzers', *TSI J. Particle Instrum.*, **3(2)** and **4(1)**, 3.

R. F. Hounam (1971), 'The Konimiser — A Dispenser for the Continuous Generation of Dust Clouds from Milligram Quantities of Asbestos', *Ann. Occup. Hyg.*, **14(4)**, 329.

M. J. Hurford (1981), 'Production of Ferric Oxide Aerosols with a May Spinning-Top Aerosol Generator', *J. Aerosol Sci.*, **12(5)**, 441.

ISO (1981), International Organization for Standards, 'Size Definitions for Particle Sampling: Recommendations of Ad Hoc Working Group Appointed by Committee TC 146 of the International Standards Organisation', *Am. Ind. Hyg. Assoc. J.*, **42(5)**, A64.

R. A. Jenkins, J. P. Mitchell and A. L. Nichols (1987), 'Monodisperse Microspheres for the Calibration of Aerosol Analysers that Operate at High Temperatures', p. 197 in 'Particle Size Analysis 1985', Ed.: P. J. Lloyd, John Wiley and Sons, Chichester, UK.

N. Karasikov and M. Krauss (1988), 'Automated Analyses of Particles Collected on Membrane Filters', paper presented at ANALYTICA, Munich, Germany, and available from GALAI Laboratories on request.

N. Karasikov and M. Krauss (1989), 'Examining the Influence of Index of Refraction on Particle Size Measurements Using a Time-of-Transition Optical Particle Sizer', p. 121 in *Filtration and Separation*, March/April.

N. Karasikov, M. Krauss and G. Barazani (1988), 'Measurement of Particles in the Range of 150–1200 μm Using a Time-of-Transition Particle Size Analyser', p. 51 in 'Particle Size Analysis 1988', Ed.: P. J. Lloyd, John Wiley and Sons, Chichester, UK.

T. Katsuta, T. Shiibashi and S. Aotani (1987), 'Particle Size Measurement of Polystyrene Particles by Transmission Electron Microscope', *J. Aerosol Res. (Japan)*, **2**, 134.

B. H. Kaye (1979), 'Characterisation of Fine Particle Systems by Utilising Diffraction Pattern Analysis', p. 626 in Proc. 2nd Europ. Symp. Particle Characterisation, Nuremberg, Germany.

P. B. Keady, F. R. Quant and G. J. Sem (1983), 'Differential Mobility Particle Sizer: A New Instrument for High-Resolution Aerosol Size Distribution Measurement Below 1 μm', *TSI Quarterly*, **IX(2)**, 3.

M. Kerker (1969), 'The Scattering of Light and Other Electromagnetic Radiation', Academic Press, New York, USA.

P. D. Kinney and D. Y. H. Pui (1995), 'Inlet Efficiency Study for the TSI Aerodynamic Particle Sizer', *Part. Part. Syst. Charact.*, **12**, 188.

E. O. Knutson and D. Sinclair (1979), 'Experience in Sampling Urban Aerosols with the Sinclair Diffusion Battery and Nucleus Counter', Proc. Advances in Particle Sampling and Measurement, Ashville, North Carolina, US Department of Commerce, USA.

E. O. Knutson and K. T. Whitby (1975), 'Accurate Measurement of Aerosol Electrical Mobility Moments', *J. Aerosol Sci.*, **6**, 453.

E. O. Knutson, K. W. Pontinen and L. W. Rees (1967), 'Coagulation in 10 μm Aerosols', *Am. Ind. Hyg. Assoc. J.*, **28**, 83.

P. Kotrappa and M. E. Light (1972), 'Design and Performance of the Lovelace Aerosol Particle Separator', *Rev. Sci. Instrum.*, **43(8)**, 1106.

Y. Kousaka, K. Okuyama, M. Shimada and K. Ohshima (1988), 'A Precise Method to Determine the Diameter of Airborne Latex Particles', *J. Aerosol Sci.*, **19**, 501.

J. Kruger and A. H. Leuschner (1978), 'A Comparative Assessment of the Behaviour of Optical Particle Counters for Aerosols', *Atmos. Environ.*, **12**, 2011.

G. Langer and A. Lieberman (1960), 'Anomalous Behaviour of Aerosol Produced by Atomization of Monodisperse Polystyrene Latex', *J. Colloid Sci.*, **15**, 357.

L. Lassen (1960), 'Ein einfacher Generator zur Erzeugung monodisperser Aerosole im Größenbereich 0.15 bis 0.70 μ (Teilchenradius)', *Z. Angew. Phys.*, **12(4)**, 157.

K. W. Lee, J. A. Gieseke and W. H. Piispanen (1985), 'Evaluation of Cyclone Performance in Different Gases', *Atmos. Environ.*, **19(6)**, 847.

K. H. Leong (1981), 'Morphology of Aerosol Particles Generated from the Evaporation of Solution Drops', *J. Aerosol Sci.*, **12(5)**, 417.

G. N. J. Lewis, J. P. Mitchell, W. D. Griffiths, D. Mark and R. S. Sokhi (1993), 'Survey of User Needs, National Measurement Infrastructure for Aerosols and Particulates in the Gas Phase/VAM14', AEA Technology Report AEA-EE-0442.

M. Lippmann (1989), 'Calibration of Air Sampling Instruments', p. 73 in 'Air Sampling Instruments for Evaluation of Atmospheric Contaminants', 7th Edn, Ed.: S. V. Hering, Am. Conf. Govern. Indust. Hygienists, Cincinnati, Ohio, USA.

M. Lippmann and R. E. Albert (1968), 'A Compact Electric-Driven Spinning Disc Generator', *Am. Ind. Hyg. Assoc. J.*, **28**, 501.

B. Y. H. Liu (1976), 'Standardization and Calibration of Aerosol Instruments', p. 39 in 'Fine Particles', Ed.: B. Y. H. Liu, Academic Press, New York, USA.

B. Y. H. Liu and D. Y. H. Pui (1974), 'A Sub-Micron Aerosol Standard and the Primary Absolute Calibration of the Condensation Nucleus Counter', *J. Colloid Interface Sci.*, **47(1)**, 155.

B. Y. H. Liu and D. Y. H. Pui (1975), 'On the Performance of the Electrical Aerosol Analyser', *J. Aerosol Sci.*, **6**, 249.

B. Y. H. Liu, R. N. Berglund and J. K. Agarwal (1974), 'Experimental Studies of Optical Particle Counters', *Atmos. Environ.*, **8**, 717.

W. W. Loo, J. M. Jacklevic and F. S. Goulding (1976), 'Dichotomous Virtual Impactors for Large-Scale Monitoring of Airborne Particulate Matter', p. 311 in 'Fine Particles', Ed.: B. Y. H. Liu, Academic Press, New York, USA.

D. A. Lundgren (1967), 'An Aerosol Sampler for Determination of Particle Concentration as a Function of Size and Time', *J. Air Pollut. Control Assoc.*, **17(4)**, 225.

J. D. McCormack and R. K. Hilliard (1980), 'Aerosol Measurement Techniques and Accuracy in the CSTF', Proc. CSNI Specialists Meeting on Nuclear Aerosols in Reactor Safety, Gatlinburg, Tennessee, USA, 15–17 April 1980, NUREG/CR-1724, 249.

W. C. McCrone and J. G. Delly (1973), 'The Particle Atlas: Volume 1, Principles and Techniques', Ann Arbor Science, Ann Arbor, Michigan, USA.

A. R. McFarland, J. B. Wedding and J. E. Cermak (1977), 'Wind Tunnel Evaluation of a Modified Andersen Impactor and an All Weather Sampler Inlet', *Atmos. Environ.*, **11**, 535.

B. A. Maguire, D. Barker and D. Wake (1973), 'Size-Selection Characteristic of the Cyclone Used in the SIMPEDS 70 Mk 2 Gravimetric Dust Sampler', *Staub-Reinhalt. der Luft*, **33(3)**, 93.

V. A. Marple (1979), 'The Aerodynamic Size Calibration of Optical Particle Counters by Inertial Impactors', p. 207 in 'Aerosol Measurement', Eds.: D. A. Lundgren, F. S. Harris, W. H. Marlow, M. Lippmann, W. E. Clarke and M. D. Durham, University of Florida Presses, Gainesville, Florida, USA.

V. A. Marple and K. L. Rubow (1976), 'Aerodynamic Particle Size Calibration of Optical Particle Counters', *J. Aerosol Sci.*, **7**, 425.

V. A. Marple, B. Y. H. Liu and K. L. Rubow (1978), 'A Dust Generator for Laboratory Use', *Am. Ind. Hyg Assoc. J.*, **39**, 26.

I. A. Marshall (1995a), 'VAM14 — Annex A: Development of Cocktail Reference Materials for Aerosol Analysis', AEA Technology Report AEA-TPD-0338.

I. A. Marshall (1995b), 'VAM14 — Annex B: (I), Development of Particle Shape Standards for Aerosol Analysis', AEA Technology Report AEA-TPD-0335.

I. A. Marshall and J. P. Mitchell (1991), 'The Behaviour of Non-Spherical Particles in a Malvern Aerosizer', UK Atomic Energy Authority Report AEA RS 5167.

I. A. Marshall, J. P. Mitchell, A. L. Nichols and A. R. Jones (1988a), 'Calibration of a Polytec HC-15 Optical Particle Analyser with Water-Droplet Aerosols', UK Atomic Energy Authority Report AEEW-R 2268.

I. A. Marshall, J. P. Mitchell, A. L. Nichols and A. Van Santen (1988b), 'Calibration Studies of a Polytec HC-15 Optical Aerosol Analyser with Water Droplets', p. 289 in Particle Size Analysis 88, Ed.: P. J. Lloyd, John Wiley and Sons, Chichester, UK.

I. A. Marshall, J. P. Mitchell and W. D. Griffiths (1990), 'The Calibration of a Timbrell Aerosol Spectrometer', *J. Aerosol Sci.*, **21(7)**, 969.

I. A. Marshall, J. P. Mitchell and W. D. Griffiths (1991), 'The Behaviour of Regular-Shaped Non-Spherical Particles in a TSI Aerodynamic Particle Sizer', *J. Aerosol Sci.*, **22(1)**, 73.

T. B. Martonen (1977), 'Aerosol Sedimentation in a Spinning Spiral Duct Centrifuge', Ph.D. Thesis, University of Rochester, USA.

E. Matijevic, M. Budnik and L. Meites (1977), 'Preparation and Mechanism of Formation of Titanium Dioxide Hydrosols of Narrow Size Distribution', *J. Colloid Interface Sci.*, **61(2)**, 302.

K. R. May (1945), 'The Cascade Impactor: An Instrument for Sampling Coarse Aerosols', *J. Sci. Instrum.*, **22**, 187.

K. R. May (1949), 'An Improved Spinning Top Homogeneous Spray Apparatus', *J. Appl. Phys.*, **20**, 932.

K. R. May (1966), 'Spinning-Top Homogeneous Aerosol Generator with Shockproof Mounting', *J. Sci. Instrum.*, **43**, 841.

K. R. May (1975), 'An Ultimate Cascade Impactor for Aerosol Assessment', *J. Aerosol Sci.*, **6**, 403.

T. T. Mercer and H. Y. Chow (1968), 'Impaction from Rectangular Jets', *J. Colloid Interface Sci.*, **27(1)**, 75.

G. Mie (1908), 'Beiträge zur Optik trüber Medien, speziell kolloidaler Metallösungen', *Ann. Physik*, **25**, 377.

J. P. Mitchell (1984), 'The Production of Aerosols from Aqueous Solutions Using the Spinning Top Generator', *J. Aerosol Sci.*, **15(1)**, 35.

J. P. Mitchell and A. L. Nichols (1988), 'Experimental Assessment and Calibration of an Inertial Spectrometer', *Aerosol Sci. Technol.*, **9(1)**, 15.

J. P. Mitchell, A. L. Nichols, A. D. Smith and K. W. Snelling (1984), 'Aerosol Particle Size Analysis Using the Stöber Spiral Duct Centrifuge', p. 345 in 'Filtration and Separation', September/October.

J. P. Mitchell, P. A. Costa and S. Waters (1988), 'An Assessment of an Andersen Mark-II Cascade Impactor', *J. Aerosol Sci.*, **19(2)**, 213.

D. C. F. Muir (1965), 'The Production of Monodisperse Aerosols by a LaMer–Sinclair Generator', *Ann. Occup. Hyg.*, **8(3)**, 233.

A. L. Nichols and B. R. Bowsher (1988), 'Chemical Characterisation of Nuclear Aerosols', *Nuclear Technol.*, **81**, 233.

G. Nicolaon, D. D. Cooke, M. Kerker and E. Matijevic (1970), 'A New Liquid Aerosol Generator', *J. Colloid Interface Sci.*, **34(4)**, 534.

NIOSH (1984), National Institute for Occupational Safety and Health, 'Manual of Analytical Methods', 3rd Edn, Vol. 1, Ed.: P. M. Eller, NIOSH Publications, Cincinnati, USA.

P. J. Nolan and L. W. Pollack (1946), 'The Calibration of a Photo-electric Nucleus Counter', *Proc. Roy. Irish Acad., Sect. A*, **51**, 9.

D. T. O'Connor (1973), 'Calibration of a Cascade Centripeter Dust Sampler', *Ann. Occup. Hyg.*, **16**, 119.

C. W. Oseen (1927), 'Latest Methods and Results in Hydrodynamics', Vol. 1 — Mathematics and Its Applications, in Monographs and Textbooks, Leipzig, Germany.

K. Philipson (1973), 'On the Production of Monodisperse Particles with a Spinning Disc', *J. Aerosol Sci.*, **4**, 51.

R. G. Pinnick, J. M. Rosen and D. J. Hofmann (1973), 'Measured Light Scattering Properties of Individual Aerosol Particles Compared to Mie Scattering Theory', *Appl. Optics*, **12(1)**, 37.

R. G. Pinnick and H. J. Auvermann (1979), 'Response Characteristics of Knollenberg Light-Scattering Aerosol Counters', *J. Aerosol Sci.*, **10**, 55.

L. W. Pollack and A. L. Metnieks (1959), 'New Calibration of Photoelectric Nucleus Counters', *Geofis. Pura. Appl.*, **43**, 285.

V. Prodi (1972), 'A Condensation Aerosol Generator for Solid Monodisperse Particles', p. 169 in 'Assessment of Airborne Particles', Eds.: T. T. Mercer, P. E. Morrow, and W. Stöber, Charles C. Thomas, Springfield, Illinois, USA.

V. Prodi, C. Melandri, G. Tarroni, T. DeZaiacomo and M. Formigniani (1979), 'An Inertial Spectrometer for Aerosol Particles', *J. Aerosol Sci.*, **10(4)**, 411.

V. Prodi, T. DeZaiacomo, D. Hochrainer and K. Spurny (1982), 'Fibre Collection and Measurement with the Inertial Spectrometer', *J. Aerosol Sci.*, **13(1)**, 49.

D. Y. H. Pui and B. Y. H. Liu (1979), 'Electrical Aerosol Analyser: Calibration and Performance', p. 384 in 'Aerosol Measurement', Eds.: D. A. Lundgren, F. S. Harris, W. H. Marlow, M. Lippmann, W. E. Clarke and M. D. Durham, University of Florida Presses, Gainesville, Florida, USA.

O. G. Raabe (1968), 'The Dilution of Monodisperse Suspensions for Aerosolization', *Am. Ind. Hyg. Assoc. J.*, **29(5)**, 439.

O. G. Raabe (1976), 'The Generation of Aerosols of Fine Particles', p. 57 in 'Fine Particles', Ed.: B. Y. H. Liu, Academic Press, New York, USA.

S. Raimondo, A. A. Haasz and B. Etkin (1979), 'The Development of a Horizontal

Elutriator: The Infrasizer Mk III', University of Toronto Institute of Aerospace Studies Report 235, Canada.

A. K. Rao and K. T. Whitby (1978a), 'Non-Ideal Collection Characteristics of Inertial Impactors, I: Single-Stage Impactors and Solid Particles', *J. Aerosol Sci.*, **9(2)**, 77.

A. K. Rao and K. T. Whitby (1978b), 'Non-Ideal Collection Characteristics of Inertial Impactors, II: Cascade Impactors', *J. Aerosol Sci.*, **9(2)**, 87.

E. Rapaport and S. Weinstock (1955), 'A Generator for Homogeneous Aerosols', *Experientia*, **11**, 363.

Rayleigh, Lord (1878), 'On the Instability of Jets', *Proc. London Math. Soc.*, **10**, 4.

P. C. Reist and W. A. Burgess (1967), 'Atomization of Aqueous Suspensions of Polystyrene Latex Particles', *J. Colloid Interface Sci.*, **24**, 271.

D. Rimberg (1979), 'Counting Efficiencies of Three Single Particle Aerosol Counters', p. 321 in 'Aerosol Measurement', Eds.: D. A. Lundgren, F. S. Harris, W. H. Marlow, M. Lippmann, W. E. Clarke and M. D. Durham, University of Florida Presses, Gainesville, Florida, USA.

C. Schegk, H. Umhauer and F. Löffler (1984), 'Measurement of Drop-Size Distributions with the Aid of a Scattered Light Particle Size Counting Analyser', *Staub-Reinhalt. der Luft*, **44(6)**, 264.

G. Scheuch and J. Heyder (1986), 'Condensation-Growth of Polydisperse Ultrafine Aerosols to Monodisperse Aerosol', p. 1057 in 'Aerosols: Formation and Reactivity', Eds.: W. O. Schikarski, H. J. Fissan, and S. K. Friedlander, Pergamon Press, Oxford, UK.

G. J. Sem (1979), 'Electrical Aerosol Analyser: Operation, Maintenance and Application', p. 400 in 'Aerosol Measurement', Eds.: D. A. Lundgren, F. S. Harris, W. H. Marlow, M. Lippmann, W. E. Clarke and M. D. Durham, University of Florida Presses, Gainesville, Florida, USA.

D. Sinclair and V. LaMer (1949), 'Light Scattering as a Measure of Particle Size in Aerosols', *Chem. Rev.*, **44**, 245.

A. D. Smith (1982), 'An Assessment of a Spiral Duct Centrifuge Using Standard and High Concentration Aerosols', UK Atomic Energy Authority Report AEEW–M 2077.

W. B. Smith, J. R. R. Wilson and D. B. Harris (1979), 'A Five-Stage Cyclone System for In-Situ Sampling', *Environ. Sci. Technol.*, **13(11)**, 1387.

W. B. Smith, D. L. Iozia and D. B. Harris (1983), 'Performance of Small Cyclones for Aerosol Sampling', in Proc. 10th Conf. European Association for Aerosol Research, *J. Aerosol Sci.*, **14(3)**, 402.

K. R. Spurny (1986), 'Physical and Chemical Characterisation of Individual Airborne Particles', Ellis Horwood Press, Chichester, UK.

W. Stahlhofen, L. Armbruster, J. Gebhart and E. Grein (1975), 'Particle Sizing of Aerosols by Single Particle Observation in a Sedimentation Cell', *Atmos. Environ.*, **9**, 851.

W. Stahlhofen, J. Gebhart, J. Heyder and B. Stuck (1979), 'Production of Monodisperse Fe_2O_3 Test Aerosols with the Aid of Centrifugal Atomisation', *Staub-Reinhalt. der Luft*, **39(3)**, 73.

W. Stöber and H. Flaschbart (1969), 'Size-Separating Precipitation of Aerosols in a Spinning Spiral Duct', *Environ. Sci. Technol.*, **3(12)**, 1280.

W. Stöber and H. Flaschbart (1971), 'High Resolution Aerodynamic Size Spectrometry of Quasi-Monodisperse Latex Spheres with a Spiral Centrifuge', *J. Aerosol Sci.*, **2**, 103.

W. Stöber, H. Flaschbart and C. Boose (1972), 'Distribution Analysis of the Aerodynamic Size and the Mass of Aerosol Particles by Means of the Spiral Centrifuge in Comparison to Other Aerosol Precipitators', *J. Colloid Interface Sci.*, **39(1)**, 109.

W. Stöber, T. B. Martonen and S. Osborne (1978), 'On the Limitations of Aerodynamical Size Separation of Dense Aerosols with the Spiral Duct Centrifuge', Chapter 12 in 'Recent Developments in Aerosol Science', Ed.: D. T. Shaw, John Wiley and Sons, New York, USA.

L. Ström (1969), 'The Generation of Monodisperse Aerosols by Means of a Disintegrated Jet of Liquid', *Rev. Sci. Instrum.*, **40(6)**, 778.

D. L. Swift (1967), 'A Study of the Size and Monodispersity of Aerosols Produced in a Sinclair–LaMer Generator', *Ann. Occup. Hyg.*, **10**, 337.

J. Swithenbank, J. M. Beer, D. S. Taylor, D. Abbot and C. G. McCreath (1977), 'A Laser Diagnostic Technique for the Measurement of Droplet and Particle Size Distribution', *Prog. Astronaut. Aeronaut.*, **53**, 421.

W. W. Szymanski and B. Y. H. Liu (1986), 'On the Sizing Accuracy of Laser Optical Particle Counters', *Part. Charact.*, **3(1)**, 1.

M. I. Tillery (1974), 'A Concentric Aerosol Spectrometer', *Am. Ind. Hyg. Assoc. J.*, **34**, 62.

V. Timbrell (1972), 'An Aerosol Spectrometer and Its Applications', Chapter 15 in 'Assessment of Airborne Particles', Eds.: T. T. Mercer, P. E. Morrow and W. Stöber, Charles C. Thomas, Springfield, Illinois, USA.

R. W. Vanderpool, A. P. Black Hall and K. L. Rubow (1984), 'Generation of Large, Solid Monodisperse Calibration Aerosols', *TSI Quarterly*, **X(1)**, 3.

N. P. Vaughan (1990), 'The Generation of Monodisperse Fibres of Caffeine', *J. Aerosol Sci.*, **21(3)**, 453.

J. H. Vincent (1989), 'Aerosol Sampling: Science and Practice', John Wiley and Sons, Chichester, UK.

W. H. Walton and W. C. Prewett (1949), 'The Production of Sprays and Mists of Uniform Drop Size by Means of Spinning Disc Type Sprayers', *Proc. Phys. Soc.*, **B62(6)**, 341.

H.-C. Wang and W. John (1987), 'Particle Density Correction for the Aerodynamic Particle Sizer', *Aerosol Sci. Technol.*, **6(2)**, 191.

J. C. F. Wang and M. A. Libkind (1982), 'Particle Collection by Cyclones at High Temperature and Pressure', Sandia National Laboratories Report SAND 82–8611.

S. C. Wang and R. C. Flagan (1990), 'Scanning Electrical Mobility Spectrometer', *Aerosol Sci. Technol.*, **13(2)**, 230.

J. B. Wedding (1975), 'Operational Characteristics of the Vibrating Orifice Aerosol Generator', *Environ. Sci. Technol.*, **9(7)**, 673.

J. B. Wedding and J. J. Stukel (1974), 'Operational Limits of Vibrating Orifice Aerosol Generator', *Environ. Sci. Technol.*, **8(5)**, 456.

K. T. Whitby and W. E. Clarke (1966), 'Electrical Aerosol Particle Counting and Size Distribution Measuring System for the 0.015 to 1 μm Size Range', *Tellus*, **18**, 573.

K. T. Whitby and R. A. Vomela (1967), 'Response of Single Particle Optical Counters to Nonideal Particles', *Environ. Sci. Technol.*, **1(10)**, 801.

K. T. Whitby, D. A. Lundgren and C. M. Peterson (1965), 'Homogeneous Aerosol Generators', *Int. J. Air Water Pollut.*, **9**, 263.

K. Willeke (1980), 'Generation of Aerosols and Facilities for Exposure Experiments', Ann Arbor Science, Ann Arbor, Michigan, USA.

K. Willeke and B. Y. H. Liu (1976), 'Single Particle Optical Counter: Principle and Application', p. 697 in 'Fine Particles', Ed.: B. Y. H. Liu, Academic Press, New York, USA.

J. C. Wilson and B. Y. H. Liu (1980), 'Aerodynamic Particle Size Measurement by Laser–Doppler Velocimetry', *J. Aerosol Sci.*, **11(2)**, 139.

B. M. Wright (1950), 'A New Dust-Feed Mechanism', *J. Sci. Instrum.*, **27**, 12.

M. L. Yeoman, B. J. Azzopardi, H. J. White, C. J. Bates and P. J. Roberts (1982), 'Optical Development and Application of a Two Colour LDA System for the Simultaneous Measurement of Particle Size and Particle Velocity', UK Atomic Energy Authority Report AERE-R 10468.

A. Zahradnicek and F. Löffler (1976), 'Eine neue Dosiervorrichtung zur Erzeugung von Aerosolen aus vorgegebenen feinkörnigen Feststoffen', *Staub-Reinhalt. der Luft*, **36(11)**, 425.

Subject Index